Modeling and Simulation in Science, Engineering and Technology

Sebastian Aniţa
Viorel Arnăutu
Vincenzo Capasso

An Introduction to Optimal Control Problems in Life Sciences and Economics

From Mathematical Models to Numerical Simulation with MATLAB®

 Birkhäuser

Sebastian Aniţa
Faculty of Mathematics
University "Al.I. Cuza" Iaşi
Bd. Carol I, 11
and
Institute of Mathematics
    "Octav Mayer" Iaşi
Bd. Carol I, 8
Romania
sanita@uaic.ro

Viorel Arnăutu
Faculty of Mathematics
University "Al.I. Cuza" Iaşi
Bd. Carol I, 11
700506 Iaşi, Romania
varnautu@uaic.ro

Vincenzo Capasso
ADAMSS (Interdisciplinary Centre
    for Advanced Applied Mathematical
    and Statistical Sciences)
and
Department of Mathematics
Università degli Studi di Milano
Via Saldini 50
20133 Milano
Italy
vincenzo.capasso@unimi.it

ISBN 978-0-8176-8097-8   e-ISBN 978-0-8176-8098-5
DOI 10.1007/978-0-8176-8098-5
Springer New York Dordrecht Heidelberg London

Library of Congress Control Number: 2010937643

Mathematics Subject Classification (2010): 49-XX, 49J15, 49J20, 49K15, 49K20, 49N25, 65-XX, 65K10, 65L05, 65L06, 68-04, 91Bxx, 92-XX

For MATLAB and Simulink product information, please contact:
The MathWorks, Inc.
3 Apple Hill Drive
Natick, MA, 01760-2098 USA
Tel: 508-647-7000
Fax: 508-647-7001
E-mail: info@mathworks.com
Web: www.mathworks.com

Printed on acid-free paper

www.birkhauser-science.com

To our families

Preface

Control theory has developed rapidly since the first papers by Pontryagin and collaborators in the late 1950s, and is now established as an important area of applied mathematics. Optimal control and optimization theory have already found their way into many areas of modeling and control in engineering, and nowadays are strongly utilized in many other fields of applied sciences, in particular biology, medicine, economics, and finance. Research activity in optimal control is seen as a source of many useful and flexible tools, such as for optimal therapies (in medicine) and strategies (in economics). The methods of optimal control theory are drawn from a varied spectrum of mathematical results, and, on the other hand, control problems provide a rich source of deep mathematical problems. The choice of applications to either life sciences or economics takes into account modern trends of treating economic problems in osmosis with biological paradigms.

A balance of theory and applications, the text features concrete examples of modeling real-world problems from biology, medicine, and economics, illustrating the power of control theory in these fields.

The aim of this book is to provide a guided tour of methods in optimal control and related computational methods for ODE and PDE models, following the entire pathway from mathematical models of real systems up to computer programs for numerical simulation. There is no pretense of being complete; the authors have chosen to avoid as much as possible technicalities that may hide the conceptual structure of the selected applications. A further important feature of the book is in the approach of "learning by doing." The primary intention of this book has been to familiarize the reader with basic results and methods of optimal control theory (Pontryagin's maximum principle and the gradient methods); it provides an elementary presentation of advanced concepts from the mathematical theory of optimal control, which are necessary in order to tackle significant and realistic problems. Proofs are produced whenever they may serve as a guide to the introduction of new concepts and methods in the framework of specific applications, otherwise explicit references to the existing literature are provided. "Working examples" are conceived to help the reader bridge those introductory examples fully developed in the text to topics of current research. They may stimulate Master's and even PhD theses projects.

The computer programs are developed and presented in MATLAB® which is a product of The MathWorks, Inc. This is a very flexible and simple programming tool for beginners, but it can also be used as a high-level one. The numerical routines and the GUI (Graphical Users Interface) are quite helpful for programming. Starting with simple programs for simple models we progress to difficult programs for complicated models. The construction of every program is carefully presented. The numerical algorithms presented here have a solid mathematical basis. One of the main goals is to lead the reader from mathematical results to subsequent MATLAB programs and corresponding numerical tests.

The volume is intended mainly as a textbook for Master's and graduate courses in the areas of mathematics, physics, engineering, computer science, biology, biotechnology, and economics. It can also aid active scientists in the above areas whenever they need to deal with optimal control problems and related computational methods for ODE and PDE models.

Chapter 1 is devoted to learning several MATLAB features by examples. A simple model from economics is presented in Section 1.1.1, and models from biology may be found in Sections 1.5 and 1.7. Chapter 2 deals with optimal control problems governed by ordinary differential equations. By Pontryagin's principle more information about the structure of optimal control is obtained. Computer programs based on mathematical results are presented. Chapter 3 is devoted to numerical approximation by the gradient method. Here we learn to calculate the gradient of the cost functional and to write a corresponding program. Chapter 4 concerns age-structured population dynamics and related optimal harvesting problems. Chapter 5 deals with some optimal control problems governed by partial differential equations of reaction–diffusion type. The last two chapters connect theory with scientific research.

Basic concepts and results from functional analysis and ordinary differential equations including Runge–Kutta methods are provided in appendices.

Matlab codes, Errata and Addenda can be found at the publisher's website: http://www.birkhauser-science.com/978-0-8176-8097-8.

We wish to thank Professor Nicola Bellomo, Editor of the Modeling and Simulation in Science, Engineering, and Technology Series, and Tom Grasso from Birkhäuser for supporting the project.

Last but not the least, we cannot forget to thank Laura-Iulia [SA], Maria [VA], and Rossana [VC], for their patience and great tolerance during the preparation of this book.

Iaşi and Milan
May, 2010

Sebastian Aniţa
Viorel Arnăutu
Vincenzo Capasso

Symbols and Notations

$I\!N$	the set of all nonnegative integers
$I\!N^*$	the set of all positive integers
$\mathbb{Z}$	the set of all integers
$I\!R$	the real line $(-\infty, +\infty)$
$I\!R^*$	$I\!R \setminus \{0\}$
$I\!R^+$ or $I\!R_+$	the half-line $[0, +\infty)$
$I\!R_+^*$	the interval $(0, +\infty)$
$I\!R^n$	the n-dimensional Euclidean space
$x \cdot y$	the dot product of vectors $x, y \in I\!R^n$
$\|\cdot\|_X$	the norm of a linear normed space X
$\nabla h, h_x, \dfrac{\partial h}{\partial x}$	the gradient of the function h
$h_x, \dfrac{\partial h}{\partial x}$	the matrix of partial derivatives of h with respect to $x = (x_1, x_2, ..., x_k)$
A^*	the adjoint of the linear operator A
$\Omega \subset I\!R^n$	an open subset of $I\!R^n$
$L^p(\Omega), 1 \le p < +\infty$	the space of all p-summable functions on Ω
$L^\infty(\Omega)$	the space of all essentially bounded functions on Ω
$L^p(0, T; X)$	(X a Banach space) the space of all p-summable functions (if $1 \le p < +\infty$), or of all essentially bounded functions (if $p = +\infty$), from $(0, T)$ to X
$L^p_{loc}(0, T; X)$	(X a Banach space) the space of all locally p-summable functions (if $1 \le p < +\infty$), or of all locally essentially bounded functions (if $p = +\infty$), from $(0, T)$ to X
$L^\infty_{loc}([0, A))$	the set of all functions from $[0, A)$ to $I\!R$ belonging to $L^\infty(0, \tilde{A})$, for any $\tilde{A} \in (0, A)$
$L^\infty_{loc}([0, A) \times [0, T])$	the set of all functions from $[0, A) \times [0, T]$ to $I\!R$ belonging to $L^\infty((0, \tilde{A}) \times (0, T))$, for any $\tilde{A} \in (0, A)$

$C([a,b];X)$ the space of all continuous functions from $[a,b]$ to X

$AC([a,b];X)$ the space of all absolutely continuous functions from $[a,b]$ to X

$C^k([a,b];X)$ the space of all functions from $[a,b]$ to X, k times differentiable, with continuous kth derivative

$N_K(u)$ the normal cone to K at u

Contents

1

An introduction to MATLAB®. Elementary models with applications

1.1 Why MATLAB®?

At the first sight, **MATLAB** (MATrix LABoratory) is a very flexible and simple programming tool. But it can also be used as high-level programming language. MATLAB is our choice because it offers some important advantages in comparison to other programming languages. This **MathWorks**^{TM} product contains a general kernel and toolboxes for specialized applications. A beginner should start with the kernel. As already mentioned, the language is easy to learn and to use, but it offers control flow statements, functions, data structures, input/output statements, and other facilities. The Mathematical Function Library provides a large set of functions for a wide range of numerical algorithms. The MATLAB GUI (Graphical User Interface) is also very good and the corresponding functions are easy to use. It is also possible to write C programs that interact with MATLAB code.

1.1.1 Arrays and matrix algebra

The basic element of MATLAB is the **matrix**. Even a simple variable is considered as a 1-by-1 matrix. The basic type is **double** (8 bytes).

Next we say a few words about the following.

The format command. To output the numerical values a standard fixed point format with four digits after the decimal point is used. It is equivalent to the command

$>> format\ short$

If we desire a longer output format we have to use other forms of the format command, such as

format long: scaled fixed point format with 15 digits
format short e: floating point format with 5 digits

S. Aniţa et al. *An Introduction to Optimal Control Problems in Life Sciences and Economics*, Modeling and Simulation in Science, Engineering and Technology, DOI 10.1007/978-0-8176-8098-5_1, © Springer Science+Business Media, LLC 2011

format long e:	floating point format with 15 digits
format short g:	best of fixed or floating point format with 5 digits
format long g:	best of fixed or floating point format with 15 digits

Look for instance to the following dialogue,

$>> pi$

$$ans =$$

$$3.1416$$

$>> format\ long$
$>> pi$

$$ans =$$

$$3.14159265358979$$

Here and in the sequel $>>$ stands for the prompter. To learn more about the format capabilities say simply

$>> help\ format$

or

$>> help\ sprintf$

sprintf allows ANSI C formatting. Let us point out that *format short* is the implicit option. Moreover, we can obtain more digits for the output by using the Variable Precision Arithmetic (vpa). For instance,

$>> vpa(pi, 100)$

will give an approximation of π with 100 digits. To learn more ask

$>> help\ vpa$

Arrays. There are no statements to declare the dimensions of an array. The simplest way to allocate a small matrix is to use an explicit list. The components on the same row should be separated by a blank or comma. The rows are separated by a semicolon or $< enter >$. The matrix itself is delimited by square brackets, that is, [ ]. For instance,

$>> A = [1\ 2\ ;\ 3\ 4]$

returns the matrix

$$A =$$

$$\begin{matrix} 1 & 2 \\ 3 & 4 \end{matrix}$$

The components of an array may be real numbers, complex numbers, or any MATLAB expression. For instance,

```
>> w = [0.7  sqrt(2)  (1 + 9) * 4/5]
```

returns the vector ($sqrt(c)$ provides a numerical approximate for $\sqrt{c}$):

$$w = [0.7000 \quad 1.4142 \quad 8.0000]$$

To get an array element we should use indexing. By

```
>> y = w(2)
```

we get

$$y = 1.4142$$

Let us point out that the first value in the implicit indexing is 1. This explains the above value of y. If a component outside the current dimensions is invoked, then the array is extended and the components not yet defined receive the value 0. For instance

```
>> w(6) = -w(1) ;
>> w
```

returns

$$w = [0.7000 \quad 1.4142 \quad 8.0000 \quad 0 \quad 0 \quad -0.7000]$$

If we invoke a variable by its name, its current value is returned (see, for instance, pi and the dialogue above). This is also true for an assignment statement; that is, the assigned variable value is returned. If we want to avoid returning the current value, which can mean many values if the variable is an array, we should place a semicolon at the end of the statement.

Let us come back to arrays. Consider the matrix A defined above and say

```
>> A(2, 4) = 7 ;
>> A
```

We therefore get

$$A =$$

$$
\begin{matrix}
1 & 2 & 0 & 0 \\
3 & 4 & 0 & 7
\end{matrix}
$$

We can also delete rows or columns of a matrix. For example, we delete the third column of the matrix A above by

```
>> A(:, 3) = [ ]
```

Then A is changed to

$$A =$$

$$
\begin{matrix}
1 & 2 & 0 \\
3 & 4 & 7
\end{matrix}
$$

We can also concatenate arrays. Consider the following statements.

```
>> A1 = [1 2 ; 3 4] ;
>> A2 = [5 6 ; 7 8] ;
>> M = [A1 ; A2]
```

Then we get the matrix

$$M =$$

$$
\begin{matrix}
1 & 2 \\
3 & 4 \\
5 & 6 \\
7 & 8
\end{matrix}
$$

For

```
>> N = [A1  A2]
```

we get the matrix

$$N =$$

$$
\begin{matrix}
1 & 2 & 5 & 6 \\
3 & 4 & 7 & 8
\end{matrix}
$$

The colon is also useful. Say

```
>> B = M(2 : 3, :)
```

Then B contains the rows 2 and 3 and all the columns of M; that is, we get

$$B =$$

$$
\begin{matrix}
3 & 4 \\
5 & 6
\end{matrix}
$$

A matrix can also be allocated by using the functions *ones*, *zeros*, and *eye*.

```
>> A = ones(m, n) ;
```

allocates A as an m-by-n matrix with all components one. Of course, the variables m and n should already have assigned values.

```
>> A = ones(n) ;
```

allocates A as an n-by-n matrix with all components one. Similar allocations are made by using *zeros*, but all components are 0. By using *eye* we can allocate the identity matrix; for example,

```
>> I = eye(10) ;
```

and I is the identity matrix 10-by-10.

To get the dimension of a matrix we use the MATLAB function *size*. Let us consider the sequence of statements:

$$>> Q \; = \; ones(4, 7) \; ;$$
$$>> a \; = \; size(Q)$$

We get

$$a \; =$$

$$4 \quad 7$$

To get the dimension of a vector we can use *size* or *length*. Let us consider the following sequence

$$>> p \; = \; [1 \; 2 \; 3 \; 4] \; ;$$
$$>> q \; = \; p' \; ;$$

Now for

$$>> length(p)$$

MATLAB returns 4 and it returns the same for

$$>> length(q)$$

For

$$>> size(p)$$

We get

$$ans \; =$$

$$1 \quad 4$$

and for

$$>> size(q)$$

it returns

$$ans \; =$$

$$4 \quad 1$$

Actually p' is the transpose of p. The function *length* can be used even for matrices. It gives $max(size)$, that is, the maximum number of rows and number of columns.

Matrix Algebra. The usual algebraic operations of matrices or of a matrix and a vector are provided. For instance, if A and B are matrices we may set

$$>> X \; = \; A + B \; ;$$

and X will store the sum of the two matrices. Of course A and B should have the same dimensions. Subtraction, $-$, and the usual product, $*$, are also available. The transpose of a matrix is indicated by $'$. For instance, $Y = X^T$ should be written as

$$>> Y \; = \; X' \; ;$$

Moreover, array-smart operations are also provided. Let $A = [a_{ij}]$ and $B = [b_{ij}]$, two matrices that have the same dimensions. Therefore

$$>> P \; = \; A.*B \; ;$$

will build a matrix $P = [p_{ij}]$ having the same dimensions and defined by

$$p_{ij} \; = \; a_{ij}b_{ij},$$

whereas

$$>> Q \; = \; A./B \; ;$$

provides the matrix $Q = [q_{ij}]$ defined by

$$q_{ij} \; = \; a_{ij}/b_{ij}.$$

By

$$>> D \; = \; A.*A$$

we get $D = [d_{ij}]$ defined by

$$d_{ij} \; = \; a_{ij}^2.$$

The Gaussian elimination algorithm is implemented to solve linear algebraic systems. If the matrix A and the right-hand side b of the system $Ax = b$ are built we get the solution by use of the operator "\"; that is,

$$>> x \; = \; A\backslash b;$$

A practical example of using Gaussian elimination is given below.

An example from economics. The problem to be solved is known as the problem of production costs (e.g., [DC72, Chapter 4.13]). A company has n sections (departments). For every section, the direct cost D_i is known. But the staff of the company would like to find the net (real) cost N_i of each section. The problem arose because a number of persons from Section j work a number of days of a given month for the benefit of another Section, k. And this occurs for each pair (j, k) of Sections, $j, k = 1, 2, \ldots, n$. Denote by T_i the total cost of Section i and we get the equations:

$$T_i = D_i + \sum_{j=1}^{i-1} P_{ji}T_j + \sum_{j=i+1}^{n} P_{ji}T_j, \quad i = 1, 2, \ldots, n.$$

Here P_{ji} is the part (percent) from the cost T_j of Section j devoted to Section i. Because the direct costs D_i, $i = 1, 2, \ldots, n$, are known, we get the linear algebraic system

$$-\sum_{j=1}^{i-1} P_{ji}T_j + T_i - \sum_{j=i+1}^{n} P_{ji}T_j = D_i, \quad i = 1, 2, \ldots, n. \tag{1.1}$$

The unknowns are the total costs T_i, $i = 1, 2, \ldots, n$. We may rewrite (1.1) as $AT = D$, where A is the system matrix, D is the right-hand side column vector, and T denotes the column vector of unknowns.

The net costs N_i, $i = 1, 2, \ldots, n$, are given by the formulae

$$N_i = P_{ii}T_i, \quad i = 1, 2, \ldots, n. \tag{1.2}$$

Therefore the problem of production costs can be solved as follows.

STEP 0: Build the matrix A and the rhs (right-hand side) D.
STEP 1: Compute the solution T (total costs) of the system $AT = D$.
STEP 2: Compute the net costs (N_i) by using Formula (1.2).

Usually the matrix A from concrete examples is invertible and well conditioned. To be more specific we consider an example from [DC72]. A company has five sections $(n = 5)$, namely

No.	Section	D_i
1	Research	2
2	Development	3
3	Production	6
4	Countability	0.5
5	IT	0.3

The monetary unit is one million USD. The matrix $[P_{ji}]$ is the following one.

$$[P_{ji}] = \begin{bmatrix} 0.6 & 0.2 & 0 & 0.1 & 0.3 \\ 0.2 & 0.6 & 0.2 & 0.2 & 0.2 \\ 0.1 & 0.2 & 0.8 & 0.5 & 0.2 \\ 0 & 0 & 0 & 0.1 & 0.3 \\ 0.1 & 0 & 0 & 0.1 & 0 \end{bmatrix}^T$$

Let us remark that the sum of every column is equal to 1 (or equivalently 100%). This shows the dispatch of the section effort. We therefore obtain (by (1.1)) the system

$$\begin{cases} T_1 - 0.2T_2 \qquad\quad - 0.1T_4 - 0.3T_5 = 2 \\ -0.2T_1 + \quad T_2 - 0.2T_3 - 0.2T_4 - 0.2T_5 = 3 \\ -0.1T_1 - 0.2T_2 + \quad T_3 - 0.5T_4 - 0.2T_5 = 6 \\ \qquad\qquad\qquad\qquad\qquad T_4 - 0.3T_5 = 0.5 \\ -0.1T_1 \qquad\qquad\qquad - 0.1T_4 + \quad T_5 = 0.3 \end{cases}$$

As concerns Formula (1.2), the values P_{ii} can be found on the diagonal of matrix $[P_{ji}]$. We therefore get

$$N_1 = 0.6T_1; \quad N_2 = 0.6T_2; \quad N_3 = 0.8T_3; \quad N_4 = 0.1T_4; \quad N_5 = 0.$$

The corresponding program follows.

```
% The production costs problem
n = 5 ;
A = eye(n) ; % load diagonal of matrix A
% load other nonzero elements
A(1,2) = -0.2; A(1,4) = -0.1; A(1,5) = -0.3;
A(2,1) = -0.2; A(2,3) = -0.2; A(2,4) = -0.2; A(2,5) = -0.2;
A(3,1) = -0.1; A(3,2) = -0.2; A(3,4) = -0.5; A(3,5) = -0.2;
A(4,5) = -0.3 ;
A(5,1) = -0.1; A(5,4) = -0.1;
A
D = [2, 3, 6, 0.5, 0.3]'
T = A\D % solve system AT = D by Gaussian elimination
P = [0.6, 0.6, 0.8, 0.1, 0]' % load values Pii
NC = P .* T % compute Net Costs
```

The percent opens a comment that is closed at the end of the corresponding line. The data of the problem are given directly in the program. Later we learn the statements *input* and *load* for a more elegant procedure.

The following table shows the differences between the direct costs D_i and the net ones N_i computed by the program above.

i	N_i	D_i
1	2.0372	2
2	3.3322	3
3	6.3593	6
4	0.0713	0.5
5	0	0.3

Let us remark that there are significant differences between the two kinds of costs for some sections.

There are other numerical functions to solve the algebraic linear system. For instance, we can replace in the above program the statement

```
T = A\D % solve system AT = D by Gaussian elimination
```

by

```
A1 = inv(A) ;
T = A1*D ;
```

The function *inv* computes the inverse of the given matrix. From the point of view of roundoff errors the Gaussian elimination behaves better. Another possibility is to use the LU-decomposition of the system matrix; that is,

```
[L,U] = lu(A) ; % L and U are triangular matrices such that A = LU
L1 = inv(L) ;
U1 = inv(U) ;
A1 = U1*L1 ;
T = A1*D
```

First A is decomposed as $A = LU$, where the matrix L is lower triangular and U is upper triangular. Next we compute $L1 = L^{-1}$ and $U1 = U^{-1}$. Because $A^{-1} = (LU)^{-1} = U^{-1} \cdot L^{-1}$, we get $A1 = U1 \cdot L1$, where $A1$ stands for A^{-1}. The numerical results are the same as above.

1.1.2 Simple 2D graphics

The plot function. If x and y are two vectors having the same length, then
$$>> plot(x, y)$$

produces a graph of y versus x. By
$$>> plot(y)$$

we get a piecewise linear graph of the components of y versus the index of these components.

We give a simple example here, the graph of the cosine function on the interval $[-\pi/2, 3\pi/2]$. We omit the prompter $>>$ below.

```
% a first graph (graph1.m)
pi2 = pi/2 ;
x = -pi2:pi/1000:3*pi2 ;
y = cos(x) ;
plot(x,y,'*') ; grid
axis([-pi2 3*pi2 -1 1])
xlabel('\bf -\pi/2 \leq {\it x} \leq 3\pi/2', 'FontSize',14)
ylabel('\bf cos(x)', 'FontSize',14)
title('\bf COS graph')
```

The short program above contains new features. A statement such as

$$x = a : h : b ;$$

builds a vector $x = [x_i]$. Its components are defined according to the rule

$$x_i = a + (i-1)h, \quad i = 1, 2, \ldots.$$

If we denote by n the length of x, then mathematically $x_n \leq b < x_{n+1}$. Of course x_{n+1} is not defined for MATLAB. The last component of the vector allocated is x_n. The statement

$$y = cos(x);$$

defines the vector $y = [y_i]$, which has the same length as x, by

$$y_i = cos(x_i), \quad i = 1, 2, \ldots, n.$$

The statement axis allows the programmer to define the limits of the axes. The first two values are the limits for the Ox-axis and the last two for the Oy-axis.

If such a statement is omitted, then the limits are established by a standard procedure. The statements *xlabel* and *ylabel* allow the programmer to write a corresponding text along each axis, respectively, by using the $Te\chi/LaTe\chi$ statements. A title can also be inserted and written above the graph. We have used '*' inside the *plot* statement to obtain a wide line on the figure for printing reasons. The figure obtained is given in Figure 1.1.

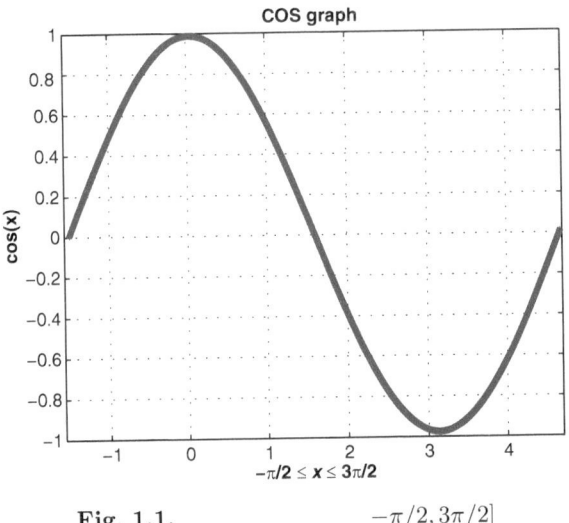

Fig. 1.1. $-\pi/2, 3\pi/2]$

1.1.3 Script files and function files

MATLAB programs. All MATLAB files have the extension **m**. Suppose that we wish to save the above program in a file graph1.m. First we say

>> edit

and the window of the text editor opens. Then we write the program in that window and we finish by choosing "save as" from the menu and giving the name of the file. That file is a **script file**. A script file is the main routine of a program (the driver). We can also make **function files**. We have already seen calls to different MATLAB functions in the previous examples. For instance *ones, zeros, eye, and plot* are such functions. Calls to different functions written by the programmer are possible in a script file and in a function file. We have to organize every function we write as a function file, which is quite simple to do. We start with an example. Our program contains a script file named grc.m and a function file named circ.m. Of course every file should be edited as explained above. The implicit name of a function file is exactly the name given in the function header.

```
% This is the script file grc.m
% It draws a circle by calling
% the function circ(x0,y0,R)
x0 = input('x for the center = ') ;
y0 = input('y for the center = ') ;
R = input('radius = ') ;
circ(x0,y0,R)
```

The *input* statement above is used to introduce data to the program. The corresponding message is written on the terminal and the program waits for the corresponding value that is assigned to the appropriate variable. The statement

```
circ(x0,y0,R)
```

is the call of the function *circ* which contains the call parameters $x0$, $y0$, R. Here we have the file circ.m

```
function circ(x0,y0,R)
% This is the function file circ.m
% The function is defined by using the parametrization of the circle
% of center = (x0,y0) and radius = R
theta = 0:pi/1000:2*pi ;
x = x0 + R*cos(theta) ;
y = y0 + R*sin(theta) ;
plot(x,y,'*') ; grid
```

The first line above is the header of the function. The statement *grid* asks for a grid to be inserted in the figure corresponding to the plot statement as can be seen in Figure 1.2. Normally a function file should have the name of the function.

Now we can launch the program grc.m in the working window by calling its name

$>> grc$

and the program will ask for the input variables and call in turn the function *circ*. Figure 1.2 is made by grc.m with the input values $x0 = 3$, $y0 = 0$, and $R = 10$.

It is also possible to call the function circ directly by giving the call parameters. For example,

$>> circ(-3, 2, 7)$

will draw a circle with center $(-3, 2)$ and radius $= 7$.

A MATLAB figure can be saved as a file with the extension *fig* by using the *save* or *save as* option from the menu. Such a file can be recovered by using the statement **figure** or the statement **open**. For instance, use

$>> figure$

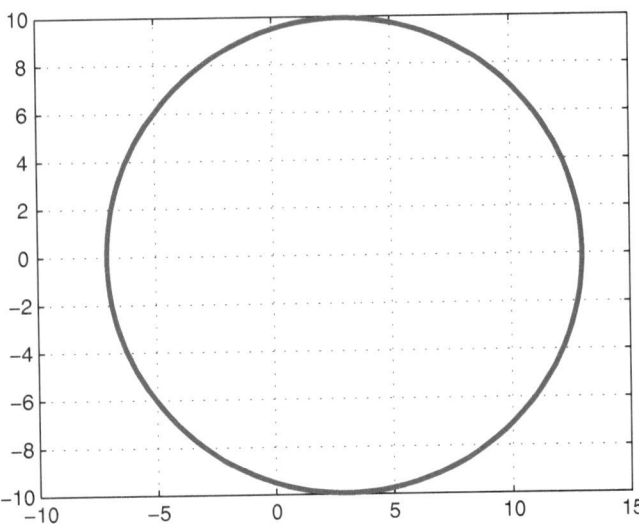

Fig. 1.2. The circle with center $(3,0)$ and radius 10

and MATLAB opens a blind figure. Then use the option *open* from the menu and give the name of the file to be restored. Another possibility is to say directly

$>>$ *open file − name.fig*

Here of course *file-name* is the name of the *fig* file. A MATLAB figure can also be exported under different file forms (extensions). One can use the option *export* from the menu and give the name and the extension of the file. A figure exported as a file with the extension *eps* can be easily processed by TeX or $LaTeX$ to be incorporated in the corresponding text.

We get back to script files to say a few things about variables inside. The variables used by a script file are stored in the memory zone (workspace) application as long as MATLAB is running. This is not true for the variables of a function file unless such a variable is declared as a global one. We give more details about global variables later. To see the variables stored in the workspace we give the command

$>>$ *whos*

If we want to delete some variables from memory we use the command *clear*

$>>$ *clear list − of − variables*

The variables from the list should be separated by a blank. If we want to remove all variables we simply say

$>>$ *clear*

We return to function files and we give the general form of the header of a function

    function [output-parameters] = function-name(input-parameters)

The input parameters are formal parameters. If a function has no input parameters the brackets ( ) are skipped. If a function has no output parameters the brackets [ ] and the = are skipped. If a function has only a variable as output parameter the brackets [ ] may be skipped. For instance, a function corresponding to $f(x) = x^2$ can be written as

    function y = f(x)
    y = x*x ;

The above function returns the value of y (no return statement is necessary).

The **help** sequence: Every function can have a help sequence. We can discover information about such a function by using the help statement; that is,

$>> help\ function - name$

We can also make a help sequence for a function we build by using the following model,

    Function Header
    % first help line
    % second help line
    % ...
    % last help line

The next line may be a void one or a statement. By using the above help statement, that is,

$>> help\ function - name$

all help lines from the first to the last will be displayed.
A function may also contain the **return** statement. The execution of such a statement ends the execution of the function. As an example we consider two ways to write the absolute-value function. It is a tutorial example because MATLAB has its own function, namely *abs*, to calculate the absolute value.

    function y = dabs(x)
    if x >= 0
        y = x ;
    else
        y = −x ;
    end

Another way to write the above function, by using the *return* statement is the following one.

```
function y = dabs(x)
if x >= 0
    y = x ;
    return
end
y = −x ;
```

The **if–else** statement has the general form

```
if condition
    block1
else
    block2
end
```

and the **if** statement has the general form

```
if condition
    block1
end
```

Here *block1* and *block2* may contain one or more statements.

I/O statements. We say a few words about input/output statements. We have already introduced the command **input** which is used to give values to some variables of a program (see the script file grc.m above). We now add some more information. Consider the statement

```
R = input('radius = ') ;
```

from grc.m. If we modify it to

```
R = input('radius = \n') ;
```

then the program writes "radius =" and skips to a new line to wait for the value. This is done by the special character "\n".
To input a string of characters we do the following.

```
name = input('What is your name? : ','s') ;
```

By using 's' we tell MATLAB to wait for a string that is assigned to the variable *name*.

To output the value of a variable we already know simply to call its identifier. For a more sophisticated output we use the statement **disp**. We give a first example:

```
name = input('What is your name? : ','s') ;
y = ['Hello ', name] ;
disp(y)
```

The program above builds the vector of characters y which is then displayed. Suppose that we give the value *John* for the variable *name*. Then the statement disp(y) will give the output

Hello John

A similar program is

```
name = input('What is your name? : ','s') ;
y = ['Hello ', name] ;
disp(' ')
disp(y)
```

To output a numerical value into a vector of characters we have to use the function **num2str**. For instance,

```
R = input('radius = ') ;
a = pi*R*R ;
x = ['surface = ', num2str(a)] ;
disp(x)
```

The **for** statement has an interesting form. We give a first example

```
for i = 1:n
    block
end
```

The inside *block* is executed for all values of i from 1 to the current value of n. We suppose of course that $n >= 1$. Another example is

```
for i = 1:2:10
    block
end
```

Here 2 is the value of the step. Therefore the corresponding values of i are 1, 3, 5, 7, 9. A feature different from other programming languages, C or FORTRAN, is that the MATLAB interpreter (compiler) builds a vector of values to be taken by the control variable. For example,

```
y = [1  4  8  20] ;
for i = y
    block
end
```

The inside *block* is executed for all values taken by i from the vector y, that is, 1, 4, 8, 20. Try the following program to check that.

```
y = [1  4  8  20] ;
for i = y
    i
end
```

Other flow-control statements, such as **while, switch, break, continue,** and **error**, are presented later when included in code examples. We also recommend the MATLAB documentation and the *help* statement.

1.2 Roots and minimum points of 1D functions

Our first goal is to discuss the approximation of a root of the equation $f(x)=0$, where f is a given real function of a real variable. The MATLAB function used is *fzero*. To be concrete let us consider

$$f(x) = x^2 - 3.$$

We therefore create the function file f.m as follows.

```
function y = f(x)
y = x*x - 3 ;
```

Then we approximate the positive root of the corresponding equation by

$\gg$ *fzero*$('f', 2)$ % or fzero(@f,2)

where 2 represents the starting point for the numerical algorithm. Another possibility is to indicate an interval where the function has a unique root. For example,

$\gg$ *fzero*$('f', [1, 2])$

In both cases a root approximation is computed.

ans =

1.7321

for $\sqrt{3}$. We can also try

$\gg x0 = $ *fzero*$('f', 2)$
$\gg f(x0)$

to see the value of f for the approximate root.
We may also use an *inline function*. Look at the following script file and notice the differences in syntax.

```
f = inline('x*x - 3') ;
x0 = fzero(f,2) % or fzero('f',2)
y = ['f(x0) = ', num2str(f(x0))] ;
disp(y)
```

Let us point out that the command *clear* also deletes the current inline functions.
To approximate the roots of polynomials we can also use the function *roots*. It suffices to give the coefficients of the polynomial by a vector. Its first component is the highest degree coefficient. For our example above we write

```
w = [1, 0, -3] ;
roots(w)
```

and we get both roots; that is,

$$ans =$$

$$1.7321$$

$$-1.7321$$

To approximate minimum points of 1D functions we can use the function *fminbnd*. Suppose that f is the inline function above; that is,

    f = inline('x*x − 3') ;

Then

$$>> fminbnd(f, -1, 1)$$

approximates the minimum point of f on the interval $[-1,1]$. Moreover,

$$>> fminbnd(f, -1, 1, 1)$$

also shows the sequence of search points of the algorithm (the last 1 above makes that option active). For more information one can use the *help* command for *fzero*, *roots*, and *fminbnd*.

    We continue with a simple example by using combined features of MATLAB. Let us consider the following equation

$$xe^x = 1 \tag{1.3}$$

on the real axis. It is of course equivalent to

$$x = e^{-x} .$$

We therefore introduce the function

$$\varphi(x) = x - e^{-x}$$

and Equation (1.3) becomes

$$\varphi(x) = 0 .$$

A simple analysis of the function φ shows that it has a unique real root located in the interval (0,1). We write down the following script file.

```
% file eqn1.m
% solve the equation
% x − exp(−x) = 0
x = 0:0.01:1 ;
y = x ;
z = exp(−x) ;
plot(x,y,'*') ; grid
hold on
plot(x,z,'r*')
hold off
fi = inline('x−exp(−x)')
```

```
pause on
pause
root = fzero(fi,0)
pause off
```

The two plot statements display on the same figure the graphs of the functions $f_1(x) = x$ and $f_2(x) = \exp(-x)$ for the interval $[0, 1]$. First the graph of f_1 is made using *plot(x,y, '*')*. The default color used is blue. The statement *hold on* asks all new plot statements to be done on the same figure until a statement *hold off* is encountered. Moreover *plot(x,z, 'r*')* is made by using the color red. To see other possibilities for colors and other features of *plot* use *help plot*. The finished figure (see Figure 1.3) allows us to approximate the root of Equation (1.3) in a graphical way.

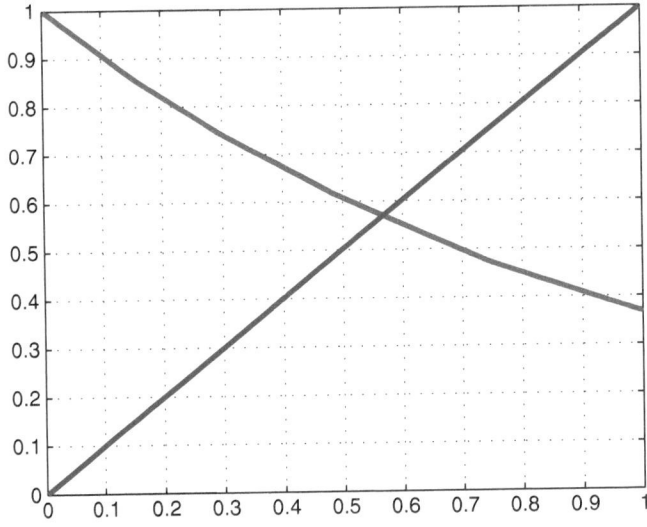

Fig. 1.3. The graphs of $y(x) = x$ and $z(x) = e^{-x}$

Next the root is approximated numerically by using the function *fzero*. The numerical approximation obtained is 0.5671. The statement *pause on* allows one to use the statement *pause* which breaks the execution of the program until a key (for instance, *enter*) is pressed. This facility is dismissed by *pause off*. It is of course possible to declare the function *fi* as a function file. We then delete the statement

```
fi = inline('x - exp(-x)')
```

from the file above and introduce the function file *fi.m*:

```
function y = fi(x)
y = x - exp(-x) ;
```

1.3 Array-smart functions

Let us consider the following simple problem of plotting the graph of $f(x) = xe^{-\sin x}/(1 + x^2)$ on the interval $[-3, 3]$. Taking into account our actual knowledge we write a function file (let us call it *fun.m*) as follows,

```
function y = fun(x)
y = x*exp(−sin(x)) / (1 + x*x) ;
```

and the script file that solves our problem

```
% file fun1.m
% Plot a graph by using the function file fun.m
clear
h = 0.001 ;
x = −3:h:3 ;
n = length(x) ;
for i = 1:n
    y(i) = fun(x(i)) ;
end
plot(x,y,'*') ; grid
```

The corresponding graph may be seen in Figure 1.4. In the above program we have obtained the vector $[y_i]_{i=1}^n$ from the vector $[x_i]_{i=1}^n$ by using *fun.m*.
A faster way to make such a transform is to build an **array-smart function**, that is, a function that computes an array from a given one applying the formula for each of its elements. We have already noticed array-smart operations when working with matrices. All usual MATLAB functions (sin, exp, etc.) are array-smart, therefore we write the following function file (let us call it *funasm.m*)

```
function y = funasm(x)
y = x.*exp(−sin(x)) ./ (1 + x.*x) ;
```

and the corresponding script file

```
% file fun2.m
% Plot a graph by using
% the function file funasm.m
clear
h = 0.01 ;
x = −3:h:3 ;
y = funasm(x) ;
plot(x,y,'*') ; grid
```

The script file below makes a detailed analysis of local minimum points and of roots of the function on the interval $[-3.3]$.

```
% file fun3.m
% test of an array-smart function
```

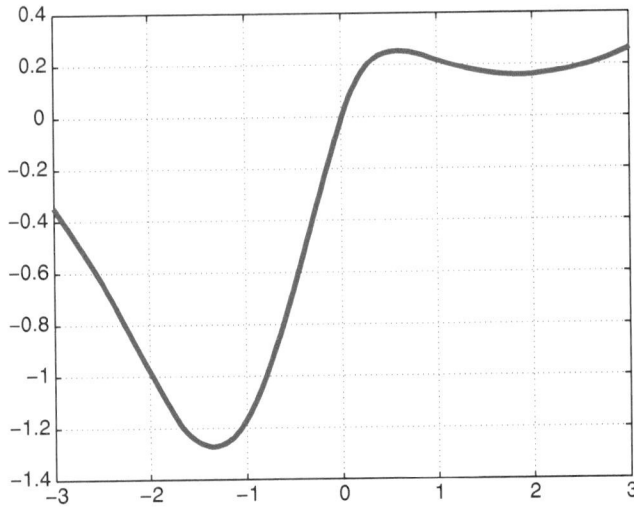

Fig. 1.4. The graph of the function $f(x) = xe^{-\sin x}/(1 + x^2)$

```
clear
h = 0.01 ;
x = -3:h:3 ;
y = funasm(x) ;
plot(x,y,'*') ; grid
pause on
pause
x1min = fminbnd('funasm',-2,-1)
pause
x2min = fminbnd('funasm',1,3)
pause
x0 = fzero('funasm',-0.5)
funx0 = funasm(x0)
pause off
```

Global variables. Each function file has its own local variables, that are separate from those of other functions, and from those of the base workspace and of script files. But if several functions, and possibly the base workspace, all declare a particular name as **global**, then they all share a single copy of that variable. Any assignment to that variable, in any file, is available to all the other files containing the same global declaration. A simple example of a *global* statement is

global a b c

Let us point out that the variables in a global list are separated by a blank. We now give an example to illustrate the use of a global variable. Let us consider the function

$$\psi(n, x) = x^n e^{-nx}$$

for $x \in [0, 10]$ and $n = 3, 4, 5, 6$. The variable x is passed to the corresponding function as a parameter, and the variable n is passed as a global variable. We therefore write the array-smart function *psi.m*

```
function y = psi(x)
global n
y = x.∧n.*exp(−n*x) ;
```

and the script file

```
% file g6.m
% plot more curves on the same figure
global n
h = 0.001 ;
x = 0:h:10 ;
n = 3 ;
plot(x,psi(x),'*') ;
grid
hold on
for n = 4:6
    plot(x,psi(x),'*') ;
end
hold off
```

The resulting graphs are given in Figure 1.5.
Another way to solve the problem above is to introduce a function, *psi1* below, which depends on two variables, namely x and n. This way global variables are no longer necessary.

```
function y = psi1(x,n)
y = x.∧n.*exp(−n*x) ;
```

The corresponding script file is

```
% file g61.m
% plot more curves on the same figure
clear
h = 0.001 ;
x = 0:h:10 ;
n = 3 ;
plot(x,psi1(x,n),'*') ; grid
hold on
for n = 4:6
```

```
    plot(x,psi1(x,n),'*') ;
end
hold off
```

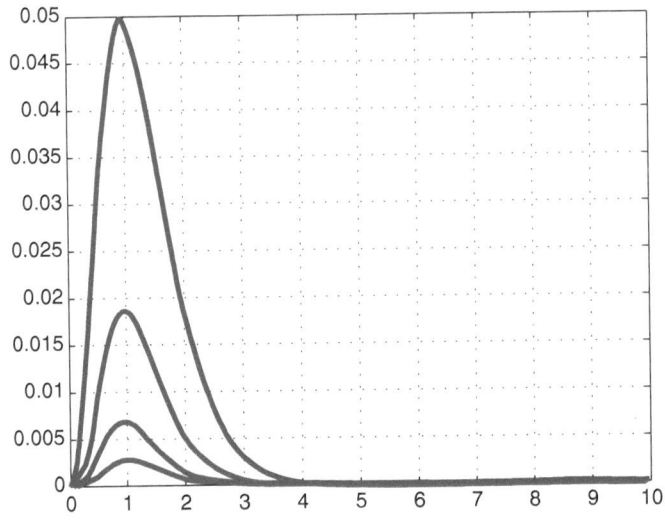

Fig. 1.5. The graphs of $\psi(n, \cdot)$, for $n \in \{3, 4, 5, 6\}$

1.4 Models with ODEs; MATLAB functions ode23 and ode45

We later introduce some simple mathematical models of ODEs (ordinary differential equations). We therefore consider the IVP (initial-value problem) for a first-order differential equation:

$$\begin{cases} y'(x) = f(x, y(x)), \\ y(x_0) = y_0, \end{cases} \tag{1.4}$$

where $y_0 \in \mathbb{R}$, and $f : D \to \mathbb{R}$ with

$$D = \{(x, y) \in \mathbb{R}^2 \; ; \; |x - x_0| \le a, \; |y - y_0| \le b\} \quad (a, \; b > 0).$$

We recall the following existence and uniqueness result (see [Har02] and [Zwi97]).

Theorem 1.1. *Assume that*

(i) f is continuous on D.

(ii) f is Lipschitz-continuous with respect to y on D; that is, there exists $L > 0$ such that

$$|f(x, y_1) - f(x, y_2)| \leq L|y_1 - y_2| \text{ for any } (x, y_1),\ (x, y_2) \in D.$$

Then there exists a unique solution of Problem (1.4), $y \in C^1([x_0 - \delta, x_0 + \delta])$, where

$$\delta = \min\{a, \frac{b}{M}\}$$

and

$$|f(x, y)| \leq M \text{ for any } (x, y) \in D, \quad M > 0.$$

Remark 1.2. If the derivative $\partial f / \partial y$ exists and is continuous on D, then the Lipschitz condition (ii) is fulfilled.

Remark 1.3. If condition (i) is replaced by $f \in C^p(D)$, where $p \in \mathbb{N}$, $p \geq 1$, then solution y belongs to $C^{p+1}([x_0 - \delta, x_0 + \delta])$.

We begin with the following particular case of (1.4).

$$\begin{cases} y'(x) = 1 - 2xy, \\ y(0) = 0. \end{cases} \tag{1.5}$$

It is obvious that the right-hand side $f(x, y) = 1 - 2xy$ of Problem (1.5) satisfies the hypotheses of Theorem 1.1 for some compact D. By using the well-known method of variation of arbitrary constants we obtain the solution of Problem (1.5) as

$$y(x) = e^{-x^2} \int_0^x e^{t^2}\, dt. \tag{1.6}$$

To exploit formula (1.6) in order to compute $y(A)$ for a given $A > 0$ we have to integrate e^{t^2} on $[0, A]$ and this can only be done numerically. Therefore, we first create a function file for the integration; let us call it *fquad*:

```
function y = fquad(t)
% function fquad to be used by prob1.m
% to solve numerically Problem (1.5)
q = t.^2 ; % array-smart
y = exp(q) ;
```

and then the script file *prob1.m*.

```
% file prob1.m
% Integrate the IVP
% y'(x) = 1 - 2xy, x in [0,A] ,
% y(0) = 0 ,
% by using the method of variation of arbitrary constants and
% the MATLAB routine quadl
```

```
clear
A = input('A : ') ;
temp1 = exp(-A*A) ;
temp2 = quadl('fquad',0,A) ;
format long
yfinal = temp1 * temp2
```

The function *quadl* numerically computes a Riemann integral (by using Simpson's formula). The call parameters are the name of the function to be integrated (a function file) and the limits of integration. An example is given above. Let us also point out that the function used by *quadl*, that is, *fquad* in our example, should be an array-smart one.

Another possibility is to integrate the IVP directly by use of a Runge–Kutta method. MATLAB provides many routines to integrate IVPs for ODEs. We start with the function file which gives the right-hand side of Problem (1.5). Let us call it *rhs2.m*.

```
function z = rhs2(x,y)
z = 1 - 2*x*y ;
```

Then we may call the corresponding function to integrate Problem (1.5). For instance

```
[x y] = ode23('rhs2', [x0 x1], y0) ;
```

The function called is *ode23* (a Runge–Kutta method of order 2–3 with adaptive step), the right-hand side of the equation is found in the file indicated by the string *'rhs2'*, the interval of integration is $[x0, x1]$, and the initial value is $y0$. The numerical results obtained are stored in the two vectors x and y. The vector $x = [x_i]_{i=1}^n$ contains the points from $[x0, x1]$ used by the adaptive step method. Due to the adaptive change of the integration step the length of x is unknown a priori but can be found after the call by *length(x)*. The vector $y = [y_i]_{i=1}^n$ contains the corresponding numerical values $y_i = y(x_i)$. It is also possible to obtain solutions (values) at specific points. For instance, suppose that $x0 = 0$ and $x1 = 1$. If we build

```
xspan = [0  0.2  0.4  0.6  0.8  1] ;
```

we may call

```
[x y] = ode23('rhs2', xspan, y0) ;
```

or

```
h = 0.01 ;
xspan = x0:h:x1 ;
[x y] = ode23('rhs2', xspan, y0) ;
```

Another MATLAB function to integrate nonstiff equations is *ode45* (a Runge–Kutta method of order 4–5 with adaptive step). Details can be obtained by the *help* statements for *ode23*, *ode45*, or *odeset*. The script file below compares the numerical results obtained by *ode23* versus *ode45* making the corresponding graphs on the same figure.

```
% file prob2.m
% Integrate the IVP
% y'(x) = 1 - 2xy, x in [x0,x1] ,
% y(0) = 0 ,
% by using the MATLAB files ode23 / ode45
% and the own function rhs2.m
clear
x0 = 0
x1 = input('x1 : ')
y0 =0
format long
[x y] = ode23('rhs2',[x0,x1],y0) ;
N = length(y)
y23 = y(N)
plot(x,y) ; grid
hold on
[z w] = ode45('rhs2',[x0,x1],y0) ;
M = length(w)
y45 = w(M)
plot(z,w,'r')
hold off
```

The values of *y23* and *y45* in the program above allow us to compare the numerical values obtained by the two functions at the final limit of the integration interval. For instance, the values obtained for the interval $[0, 3]$ are

$N = 22$; y23 $= 0.17794836747613$
$M = 61$; y45 $= 0.17829834037441$

These values can be compared with the one obtained by *prob1.m* which is

yfinal $= 0.17827103166817$

To obtain a better graphical image in the above program we have introduced the following sequence

```
h = 0.01 ;
xspan = x0:h:x1 ;
```

and we have replaced the calls to *ode* routines by

```
[x y] = ode23('rhs2',xspan,y0) ;
[z w] = ode45('rhs2',xspan,y0) ;
```

and the *plot* statements by

```
plot(x,y,'*') ; grid
plot(z,w,'r*')
```

to obtain wide lines for *plot* (more points and '*'). The corresponding graphs are given in Figure 1.6. Let us remark that the two graphs are almost identical.

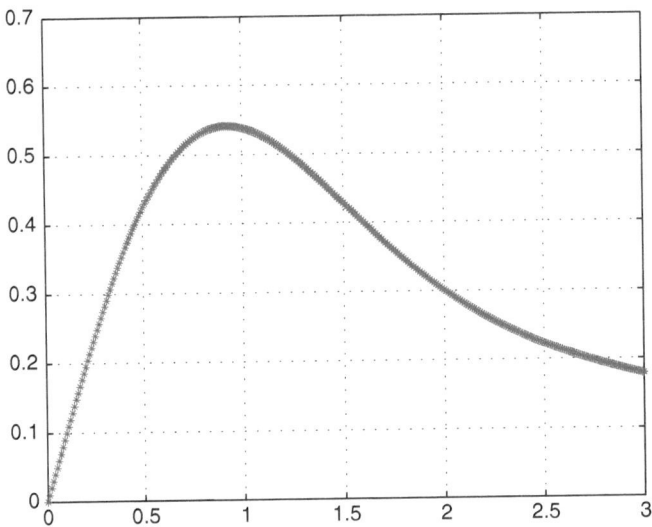

Fig. 1.6. The graphs corresponding to ode23 and ode45 are almost identical

Let us consider now another example of IVP, namely

$$\begin{cases} y'(x) = y - \dfrac{2x}{y}, \\ y(0) = 1. \end{cases} \tag{1.7}$$

We multiply the equation by y and we get

$$yy' = y^2 - 2x. \tag{1.8}$$

We introduce the new function $z = y^2$ and we get from (1.8) the equation

$$z' = 2z - 4x.$$

The corresponding initial condition is $z(0) = 1$. Therefore we readily get $z(x) = 2x + 1$ and hence the solution of Problem (1.7),

$$y(x) = \sqrt{2x + 1}. \tag{1.9}$$

We are going now to compare the mathematical solution (1.9) with the numerical one obtained by *ode23*. We first make the function file *rhs3* for the right-hand side of Problem (1.7):

    function z = rhs3(x,y)

$z = y - 2*x/y$;

The script *prob3.m* makes the comparison by representing the two graphs on the same figure.

```
% file prob3.m
% Integrates numerically the IVP (1.7)
% y'(x) = y - 2x/y , x in [0,L]
% y(0) = 1
% by using of ode23 and compares on [0,L] with
% the mathematical solution y(x) = sqrt(2x + 1)
clear
L = input('L : ') ;
x = 0:0.01:L ;
y = sqrt(2*x + 1) ;
plot(x,y,'*')
hold on
[z w] = ode23('rhs3',[0 L], 1) ;
plot(z,w,'r*')
hold off
```

For the interval $[0, L]$ with $L = 2$ the two solutions are represented in Figure 1.7.

Let us remark that the two graphs are almost identical, which means that the numerical routine behaves quite well. If we take a longer interval, namely $L = 4$, we can see the result given by Figure 1.8.

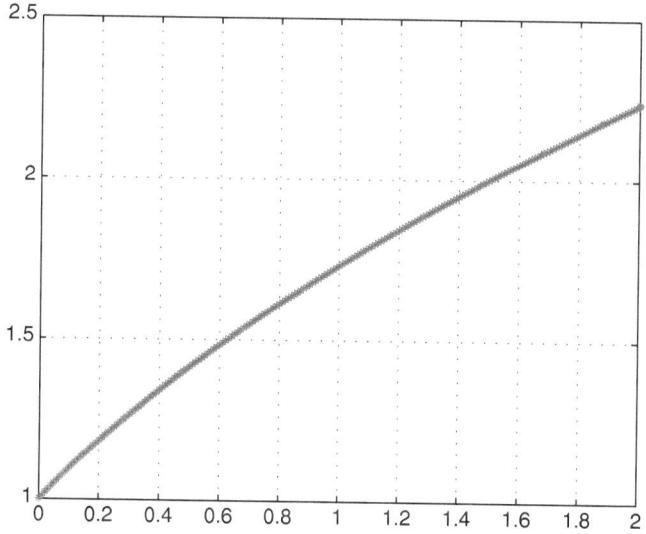

Fig. 1.7. The mathematical and numerical solutions of (1.7) on $[0, 2]$

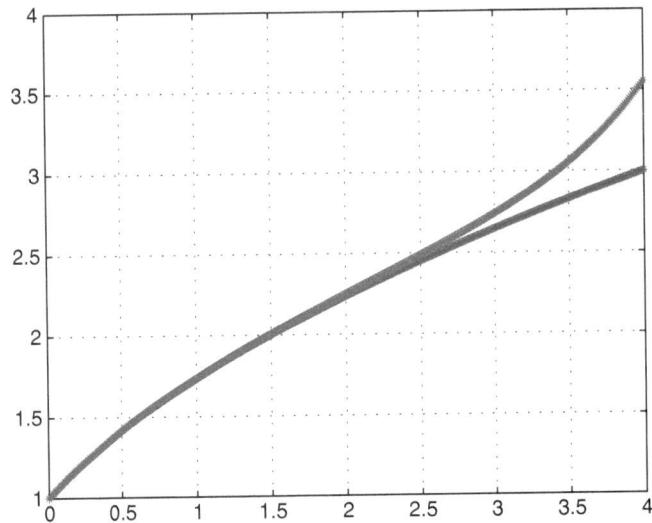

Fig. 1.8. The mathematical and numerical solutions of (1.7) on $[0, 4]$

It is quite clear that far from 0, the starting point of the integration interval, the numerical solution fails because the roundoff errors accumulate. A better numerical result is obtained replacing *ode23* with *ode45* in *prob3.m*. Such an experience is instructive and very easy to do.

Let us also point out that the function files made for the right-hand sides of IVPs as in our examples are usual functions and not array-smart ones due to the fact that the numerical methods build the solution point after point, that is, not in a vectorized way.

The above-mentioned integration methods may be efficiently used even if the right-hand side functions of the IVPs satisfy only weaker regularity properties than those in Theorem 1.1. We often apply these methods for piecewise continuous functions with excellent results.

1.5 The spruce budworm model

As a first biological model leading to an IVP for ODEs we consider *the spruce budworm model* (e.g., [Ist05, Chapter 2.3] and [All07]). The spruce budworm is an insect that damages forests in North America; it feeds on needles of coniferous trees. Let $N(t)$ be the number of individuals of the spruce budworm population at the moment $t \in [0, T]$. Its evolution can be described by a logistic model:

$$N'(t) = rN(t)\left(1 - \frac{N(t)}{K}\right),$$

which means that the population evolution depends not only on the natural fertility and mortality rates but also on a mortality rate induced by overpopulation. In the above formula, r and K are dimensionless parameters involving real field ones, namely r is the intrinsic growth rate and K is the carrying capacity of the environment. To the additional mortality rate due to overpopulation, $(r/K)N(t)$, we add now a predation term. The predator is represented by birds. The new equation is

$$N'(t) = rN(t)\left(1 - \frac{N(t)}{K}\right) - p(N(t)), \quad t \in [0, T],$$

where

$$p(N) = \frac{BN^2}{A^2 + N^2}.$$

The shape of the graph of p above, for $A = 2$ and $B = 1.5$, on the interval $[0, 20]$, is given in Figure 1.9.

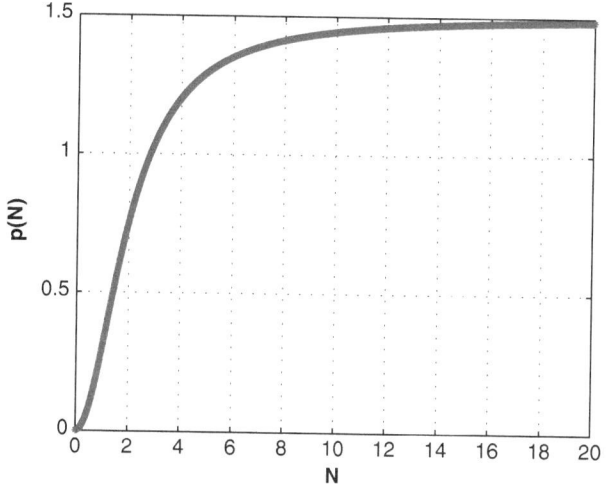

Fig. 1.9. The graph of $p(N)$ on $[0, 20]$, for $A = 2$, $B = 1.5$

Exercise. Write a program that plots the graph of the function p.

We add an initial condition and we get the IVP:

$$\begin{cases} N'(t) = rN(t)\left(1 - \frac{N(t)}{K}\right) - \frac{BN(t)^2}{A^2 + N(t)^2}, & t \in [0, T] \\ N(0) = N_0. \end{cases}$$

The right-hand side of the equation is given by the function file

```
function z = sbw1(x,y)
global r K A2 B
y2 = y*y ;
z = r*y*(1 - y/K) - B*y2 / (A2 + y2) ;
```

The corresponding script file is

```
% program ist1.m for the spruce budworm model
global r K A2 B
r = input('r = ') ;
K = input('K = ') ;
A = input('A = ') ;
A2 = A*A ;
B = input('B = ') ;
T = input('T = ') ;
N0 = input('N(0) = ') ;
h = 0.01 ;
tspan = 0:h:T ;
[x y] = ode45('sbw1', tspan, N0) ;
plot(x,y,'r*') ; grid
xlabel('\bf t','FontSize',16)
ylabel('\bf N(t)','FontSize',16)
```

The evolution of the spruce budworm population is given in Figures 1.10 and 1.11 for different choices of r, K, A, B, T and N_0.

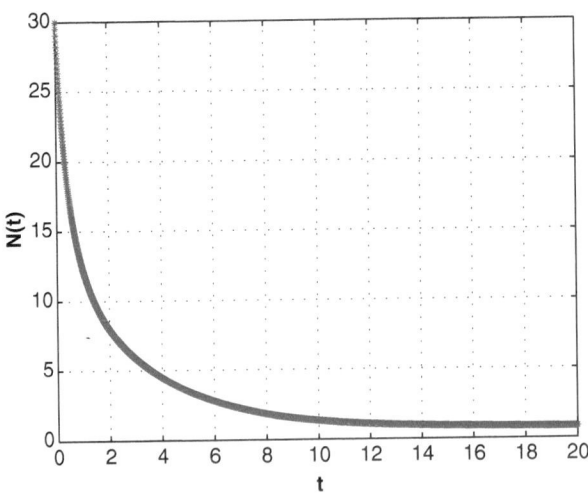

Fig. 1.10. The graph of $N(t)$, for $r = 0.3$, $K = 5$, $A = 2$, $B = 1.5$, $T = 20$, $N_0 = 30$

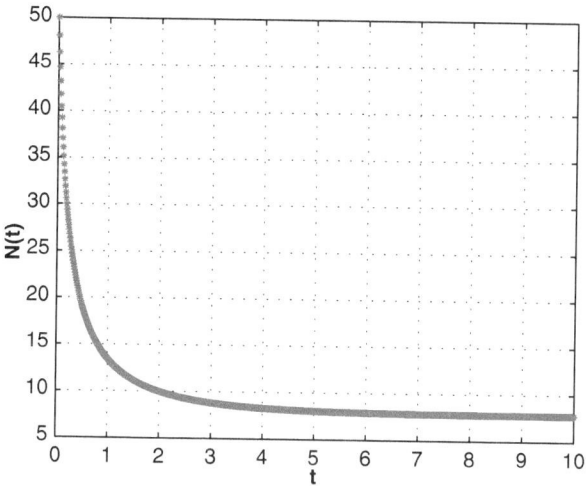

Fig. 1.11. The graph of $N(t)$, for $r = 1$, $K = 10$, $A = 3$, $B = 2$, $T = 10$, $N_0 = 50$

Exercise. Consider the following IVP, which gives the evolution of an insect population (such as the spruce budworm model) subject to a harvesting process:

$$\begin{cases} N'(t) = rN(t)\left(1 - \dfrac{N(t)}{K}\right) - u(t)N(t), & t \in [0, T] \\ N(0) = N_0. \end{cases}$$

Take, as for the first test of the spruce budworm model, $r = 0.3$, $K = 5$, $T = 20$, and $N_0 = 30$. Plot the graphs of the insect population $N(t)$ for the following choices of the harvesting effort $u(t)$.

(i) $u(t) = 0.5$ on $[0, T]$.
(ii) $u(t) = 1$ on $[0, T]$.
(iii) $u(t) = 3$ on $[0, T]$.

In all cases also compute the harvest given by $\int_0^T u(t)N(t)dt$.

Hint. We change the *global* statement from program ist1.m to

global r K u

and we also introduce the input statement

u = input('u = ') ;

We also have to modify the rhs-function for the *ode45* call. Suppose we call it *sbw2*. Thus

```
function z = sbw2(x,y)
global r K u
z = r*y*(1 − y/K) − u*y ;
```

To compute the harvest value we add the statement

```
[x y] = ode45('sbw2', [0 T], N0) ;
...
harv = u * trapz(x,y)
```

Learn more about the numerical integration function *trapz* by using *help trapz*. The harvest values obtained are presented in the following table.

u	harvest
0.5	19.0655
1	21.2371
3	25.5577

1.6 Programming Runge–Kutta methods

For the mathematics of Runge–Kutta methods we refer to Appendix A.4. We write a function to implement the standard Runge–Kutta method of order 4 with fixed step. The corresponding file is *RK4.m*.

```
function ret = RK4(y,x)
% RK4 standard method with fixed step h
% the rhs of the equation is given by zeta.m
global h
xm = x + h/2 ;
k1 = h * zeta(x,y) ;
k2 = h * zeta(xm,y + k1/2) ;
k3 = h * zeta(xm,y + k2/2) ;
k4 = h * zeta(x + h,y + k3) ;
ret = y + (k1 + k4 + 2.0*(k2 + k3))/6.0 ;
```

We consider the IVP

$$\begin{cases} y'(x) = x^2 + y^2, \\ y(0) = 0. \end{cases} \qquad (1.10)$$

We compare the numerical results obtained by *RK4* and by *ode45*. First we write the file *zeta.m* for the right-hand side of the equation

```
function z = zeta(x,y)
z = x*x + y*y ;
```

and then the script *testode1.m*. Problem (1.10) is integrated on the interval $[0, L]$.

```
% file testode1.m
clear
global h
L = input('L : ') ;
x0 = 0 ;
```

```
y0 = 0 ;
[t,w] = ode45('zeta',[x0 L],y0) ;
N = input('N : ') ; % number of subintervals for RK4
h = (L − x0) / N
for i = 1:N + 1
    x(i) = x0 + (i − 1)*h ;
end
y(1) = y0 ;
for i = 1:N
    y(i + 1) = RK4(y(i),x(i)) ;
end
format long
yrk4 = y(N + 1) % y(L) approximated by RK4
M = length(w)
y45 = w(M) % y(L) approximated by ode45
plot(t,w,'*') ; grid
hold on
plot(x,y,'r*')
hold off
```

The numerical results are compared graphically and by the approximated value of $y(L)$, where $L = 1$. The two graphs are given in Figure 1.12 showing no difference.

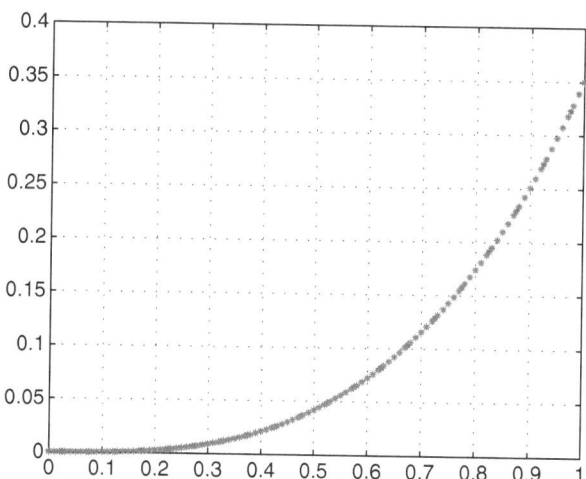

Fig. 1.12. The graphs of numerical solutions obtained by RK4 and by ode45

The result obtained by *ode45* was $y(L) = 0.35023184134841$ for $M = 41$. Here $M − 1$ is the number of subintervals. For *RK4* we have obtained 0.35023184453449 for $N = 100$ and 0.35023184431678 for $N = 1000$, where

N is the number of subintervals. The results obtained by $RK4$ are good enough compared to that obtained by $ode45$ because Problem (1.10) is nonstiff.

1.7 Systems of ODEs. Models from Life Sciences

We now treat biological models that lead to systems of ODEs with n equations and n unknown functions, for $n = 2, 3, 4$.

A predator–prey model

Predator–prey models can be found, for instance, in [BC98, Section 5.4], [Smo83, Chapter 14], [http1]. Here we consider a model with prey overcrowding. That is, the predator's appetite is satiated as the prey population increases. We denote the satiable predator population at time t by $y_1(t)$ and the corresponding prey population by $y_2(t)$. The dynamics of this predator–prey system is described by the following ODE system:

$$\begin{cases} y_1' = -ay_1 + \dfrac{by_2}{c + ky_2}y_1, \\ y_2' = (d - ey_2)y_2 - \dfrac{fy_2}{c + ky_2}y_1, \end{cases} \tag{1.11}$$

for $t \in [0, L]$. Here a, b, c, d, e, f are nonnegative constants. The positive parameter k measures the predator's satiation threshold. A small k means that it takes a lot of prey before there is any satiation. A large k means that the satiation quickly appears as the number of prey increases.

For our numerical tests we have considered (see the reference above) $a = 0.5$, $b = d = e = f = 1$, $c = 0.3$, $k = 0.7$. The final time is $L = 200$ and the initial conditions are

$$y_1(0) = 0.5, \quad y_2(0) = 1. \tag{1.12}$$

The program to solve the IVP (1.11) and (1.12) follows.

```
% Satiable Predation model – sp1.m
% http://www.math.hmc.edu/resources/odes/odearchitect/examples/
clear
global a b c d e f k
load file1.txt
disp('get model parameters') ;
a = file1(1) ;
b = file1(2) ;
c = file1(3) ;
d = file1(4) ;
e = file1(5) ;
f = file1(6) ;
k = file1(7) ;
```

```
disp('get data') ;
L = input('final time : ') ;
y01 = input('y1(0) : ') ;
y02 = input('y2(0) : ') ;
lw = input('LineWidth : ') ; % for graphical use (plot)
tspan = 0:0.01:L ;
[t y] = ode45('sprhs1',tspan,[y01 ; y02]) ;
% graphs
plot(t,y(:,1),'LineWidth',lw) ; grid
hold on
plot(t,y(:,2),'r','LineWidth',lw)
legend('predator','prey',0)
hold off
```

Let us first point out that the values of the parameters are loaded from the file *file1.txt*. In that file the values are written one by row. The file can be made by using the MATLAB editor or any other text editor of the OS (Operating System). The statement

    load file1.txt

provides a vector called file1 which contains the corresponding values. Those values are transferred to the corresponding variables by the statements that follow

    a = file1(1) ;
    ...

Next we see that there are some changes when solving a system of two equations in comparison to the case of one. Let us first see the parameters of *ode45*. For the name of the right-hand side file function nothing changes. The interval of integration, in our case $[0, L]$, is also given the same way. A change appears for the initial conditions because we have to give two values instead of one. The syntax can be observed above, that is, $[y01; y02]$. The results returned by *ode45* also look different. The vector $t = [t_i]_{i=1}^n$ contains the points from $[0, L]$ used by the adaptive step method, whereas y is a matrix with n rows and two columns, where $n = length(t)$. The first column contains the corresponding numerical values of y_1 and the other column of y_2. So on the same figure our *plot* statement constructs the graphs of $y_1(t)$, the predator population, and of $y_2(t)$, the prey population.

The statement *legend* produces a legend for the figure. The text is written according to the order of the *plot* statements executed on the same figure. The parameter 0 is related to the location of the legend on the figure. The choices are:

    0 = Automatic "best" placement (least conflict with data)
    1 = Upper right-hand corner (default)

2 = Upper left-hand corner

3 = Lower left-hand corner

4 = Lower right-hand corner

−1 = To the right of the plot

For more information use *help legend*.

Let us now explain the presence of the vector *tspan* which contains time-grid points. Its role is to force *ode45* to use these grid points for the integration points. Therefore the number of points (components) of the resulting vector t increases and the graphs obtained "look better." Moreover, the *plot* statements contain the parameter 'LineWidth' which allows us to establish the width of the drawing line. The corresponding value may be given directly by a constant or by a variable (as *lw* in our program). For Figures 1.13 and 1.15 below we have used *lw*=3. Also try other values.

The function corresponding to the rhs of the system is given below.

```
function out1 = sprhs1(t,y)
global a b c d e f k
out1 = [ −a*y(1) + b*y(1)*y(2)/(c + k*y(2)) ; ...
(d − e*y(2))*y(2) − f*y(1)*y(2)/(c + k*y(2)) ] ;
```

Let us explain how to write the value assigned to the return variable, namely *out1* above. We have to introduce the right-hand side (rhs) for both equations of the system. They are put into [ ] and are separated by ";". y_1 is written as $y(1)$ and y_2 as $y(2)$.

The graphs are given in Figure 1.13. To make a more explicit graph we have added *sp1.m* text statements at the end of program:

```
text(100,0.65,'\bf predator','FontSize',14)
text(100,0.05,'\bf prey','FontSize',14)
```

We cite from the MATLAB help:

"text(X,Y,'string') adds the text in the quotes to location (X,Y) on the current axes, where (X,Y) is in units from the current plot".

For more information try *help text*.

Remark that both populations are stabilizing to periodic functions.

To get a graphical representation in the (y_1, y_2) plane we change the "graphs" part of program *sp1.m* to

```
...
% graphs
plot(y(:,1),y(:,2),'*') ; grid
axis([0 1 0 1])
xlabel('\bf y_1','FontSize',16) ;
ylabel('\bf y_2','FontSize',16) ;
```

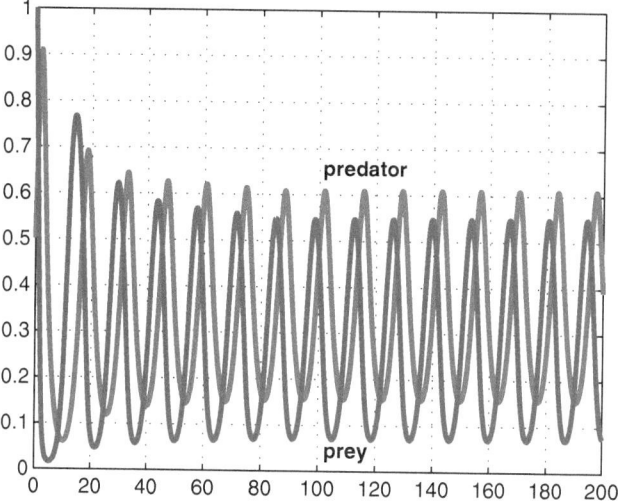

Fig. 1.13. The distribution of predators and prey

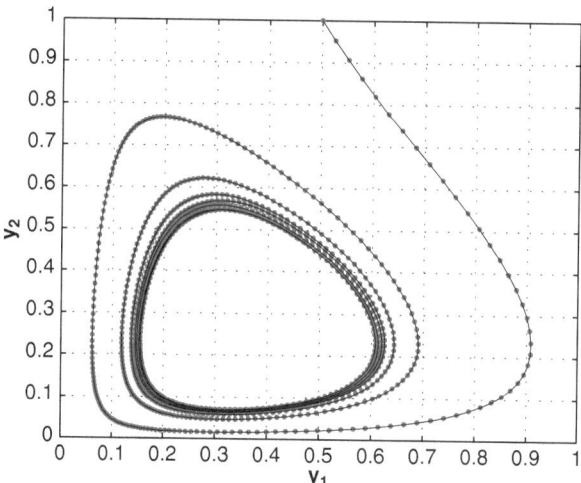

Fig. 1.14. Representation of predators and prey in (y_1, y_2) coordinates

The corresponding graph is given by Figure 1.14.

If we replace the initial conditions (1.12) by

$$y_1(0) = 0.42, \quad y_2(0) = 0.3, \tag{1.13}$$

the graphs of the components of the solution to the IVP (1.11), (1.13) are given in Figure 1.15. The corresponding (y_1, y_2) graph is given in Figure 1.16. We have changed the *axis* statement to get a focused picture.

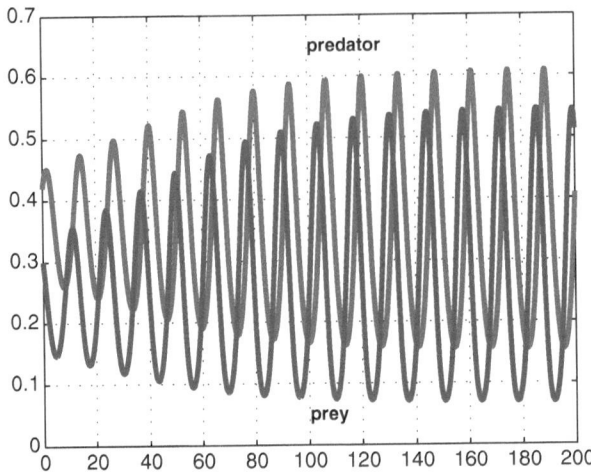

Fig. 1.15. The distribution of predators and prey for $y_1(0) = 0.42$, $y_2(0) = 0.3$

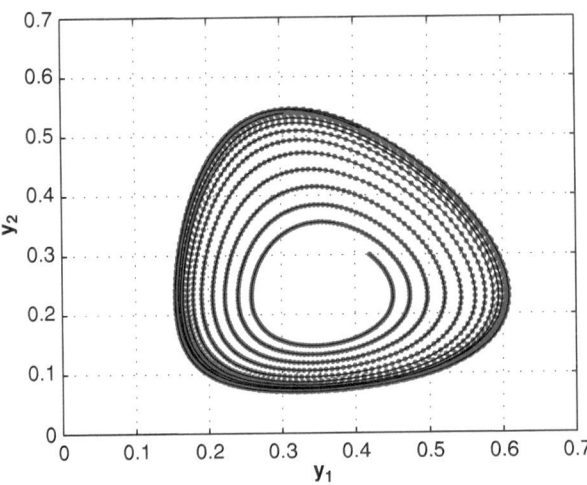

Fig. 1.16. Representation of predators and prey in (y_1, y_2) coordinates for $y_1(0) = 0.42$, $y_2(0) = 0.3$

The Fitzhugh–Nagumo Equations. A model for neural activity

The Fitzhugh–Nagumo equations represent a model for electrical activity in a neuron (e.g., [Mur89, Section 6.5], [Smo83, Chapter 14], and [http1]). The neuron is an excitable system that can be stimulated with an input, such as an electric current. The state of the excitation is described by the function y_1, which represents the voltage in the neuron as a function of time. When a neuron is excited, physiological processes will cause it to recover from the excitation. The recovery is represented by the function y_2. We start with the Fitzhugh–Nagumo system:

$$\begin{cases} y_1' = c\left(y_1 + y_2 - \dfrac{y_1^3}{3}\right), \\ y_2' = c^{-1}(a - y_1 - by_2), \end{cases}$$

for $t \in [0, L]$. Two kinds of behavior can be observed in real neurons:

- The response y_1 of the neuron tends to a steady state after a large displacement; the neuron has **fired**; it is a single **action-potential**;
- The response y_1 is a periodic function; the neuron experiences **repetitive firing**.

The parameters a, b, and c should satisfy the following constraints for meaningful behavior:

$$1 - \frac{2}{3}b < a < 1, \quad 0 < b < 1, \quad b < c^2.$$

For the numerical tests we take $a = 0.75$, $b = 0.5$, $c = 1$. With $L = 20$ and the initial conditions

$$y_1(0) = 3, \quad y_2(0) = 0, \tag{1.14}$$

we get the graphs in Figure 1.17.

It is clear that the neuron is fired; that is, y_1 tends to a steady state. The program is similar to *sp1.m*. We give only the rhs-function

```
function out1 = fitzrhs1(t,y)
global a b c
out1 = [ c*(y(2) + y(1) - y(1)∧3/3) ; (a - b*y(2) - y(1))/c ] ;
```

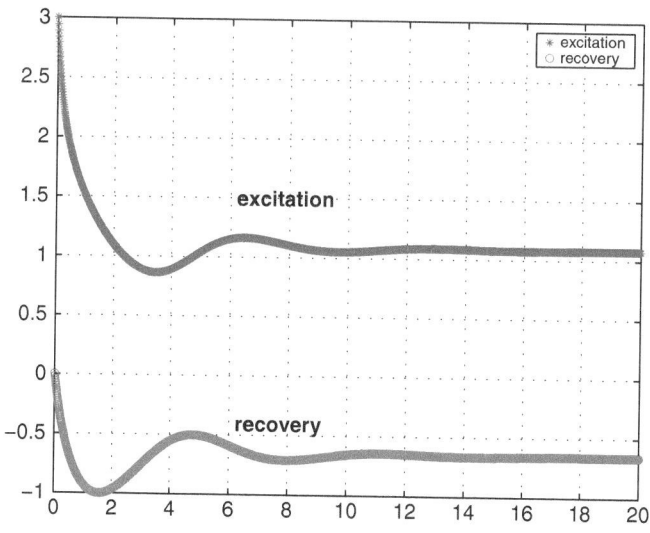

Fig. 1.17. The graphs of the excitation and recovery functions

We now introduce a **stimulus** in the model. It is represented by the time-dependent function z.

$$\begin{cases} y_1' = c\left(y_1 + y_2 - \dfrac{y_1^3}{3} \right) - z, \\ y_2' = c^{-1}(a - y_1 - by_2), \end{cases}$$

for $t \in [0, L]$. For the following numerical test we take

$$z(t) = \begin{cases} 0, & \text{for } t \in [0, t^*] \\ v, & \text{for } t \in (t^*, L] \end{cases} \tag{1.15}$$

where $0 < t^* < L$ is a switching point and $0 < v < 1$ is the stimulus value. The corresponding program follows.

```
% the Fitzhugh–Nagumo equation – fitz2.m
% z = the stimulus to the neuron
% http://www.math.hmc.edu/resources/odes/odearchitect/examples/
clear
global a b c
global v tsw
load file2.txt
disp('get model parameters') ;
a = file2(1) ;
b = file2(2) ;
c = file2(3) ;
disp('get data') ;
L = input('final time : ') ;
y01 = input('y1(0) : ') ;
y02 = input('y2(0) : ') ;
v = input('stimulus : ') ; % stimulus value
tsw = input('switch time : ') ; % switch time for the stimulus
tspan = 0:0.01:L ;
[t y] = ode45('fitzrhs2',tspan,[y01 ; y02]) ;
plot(t,y(:,1),'*',t,y(:,2),'ro') ; grid
legend('excitation','recovery',0)
text(40,1.5,'\bf excitation','FontSize',16)
text(40,-0.5,'\bf recovery','FontSize',16)
n = length(tspan) ;
for i = 1:n
    if tspan(i) > tsw
        z(i) = v ;
    else
        z(i) = 0 ;
    end
end
```

```
figure(2)
plot(tspan,z,'*') ; grid
axis([0 L 0 v])
xlabel('\bf t','FontSize',16)
ylabel('\bf stimulus z(t)','FontSize',16)
```

The corresponding rhs-function is

```
function out1 = fitzrhs2(t,y)
global a b c
global v tsw
if t > tsw
    z = v ;
else
    z = 0 ;
end
out1 = [ c*(y(2) + y(1) − y(1)∧3/3) − z ; (a − b*y(2) − y(1))/c ] ;
```

Let us point out that in *fitzrhs2* we have used the variable t, which stores the current time value, to compute the current stimulus z. For the numerical test we have considered the parameters a, b, c as before, $L = 100$, $v = 0.5$, and $t^* = 30$ (the program variable is *tsw*) in formula (1.15) and the initial conditions (1.14). Excitation y_1 and recovery y_2 corresponding to the stimulus z are given in Figure 1.18.

As for the prey–predator model we also have the corresponding (y_1, y_2)-graph which is given in Figure 1.19. We have managed the *axis* statement to get a focused picture.

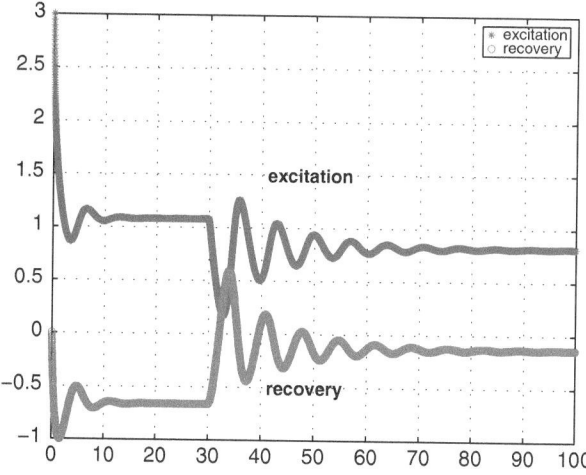

Fig. 1.18. The graphs of the excitation and recovery functions when stimulus occurs

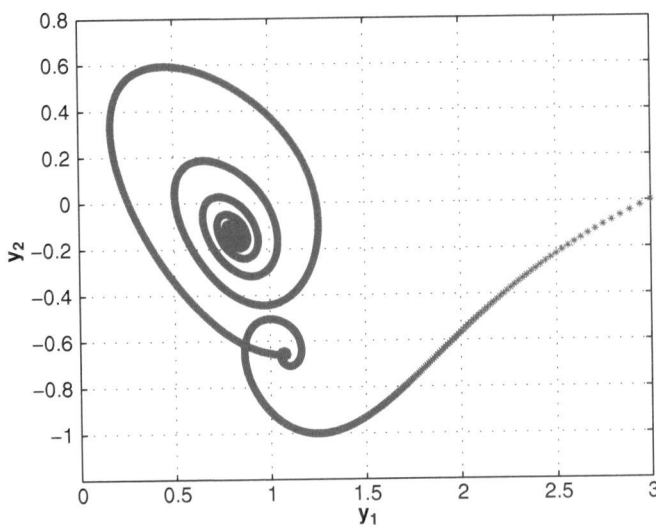

Fig. 1.19. The excitation and recovery (with stimulus) in (y_1, y_2)-coordinates

We now consider a 3×3 model, that is, a model with three equations and three unknown functions.

Lead in the Body

Lead is a toxic element that is present in different products. The model is a compartmental one for the transport of lead through the body (e.g., [BC98, Section 7.1] and [http1]). We consider three compartments: **blood** (compartment 1), **tissues** (compartment 2), and **bones** (compartment 3). Denote by I_1 the lead input rate into blood and by $y_i(t)$, $i = 1, 2, 3$, the amount of lead in compartment i at time t. There are exchanges of lead between blood and tissues and between blood and bones. The corresponding exchange rates are k_{21} and k_{12} for blood and tissues and k_{31} and k_{13} for blood and bones, respectively. The lead also passes from blood to urine (exchange rate k_{01}) and from tissues to hair, nails, and sweat (exchange rate k_{02}). Therefore the differential system is

$$\begin{cases} y_1' = -(k_{01} + k_{21} + k_{31})y_1 + k_{12}y_2 + k_{13}y_3 + I_1, \\ y_2' = k_{21}y_1 - (k_{02} + k_{12})y_2, \\ y_3' = k_{31}y_1 - k_{13}y_3, \end{cases} \qquad (1.16)$$

for $t \in [0, L]$. We add the initial conditions

$$y_1(0) = y_2(0) = y_3(0) = 0, \qquad (1.17)$$

which mean that there is no lead in the body at time $t = 0$. The program to integrate and to make a subsequent graphical representation follows.

```
% Lead in the Body – lead1.m
% y(1) = lead in Blood
% y(2) = lead in Tissues
% y(3) = lead in Bones
% http://www.math.hmc.edu/resources/odes/odearchitect/examples/
clear
global k01 k21 k31
global k02 k12 k13
global I1
load file3.txt
disp('get model parameters') ;
k01 = file3(1) ;
k21 = file3(2) ;
k31 = file3(3) ;
k02 = file3(4) ;
k12 = file3(5) ;
k13 = file3(6) ;
I1 = file3(7) ;
disp('get data') ;
L = input('final time : ') ;
y01 = input('y1(0) = ') ;
y02 = input('y2(0) = ') ;
y03 = input('y3(0) = ') ;
tspan = 0:0.01:L ;
[t y] = ode45('lrhs1',tspan,[y01 ; y02 ; y03]) ;
plot(t,y(:,1),'ro',t,y(:,2),'*',t,y(:,3),'gs') ; grid
% legend('blood','tissues','bone',0)
text(400,400,'tissues')
text(400,1400,'blood')
text(400,2000,'bones')
```

Because the system is 3×3 there are changes in comparison to a 2×2 system. In the call to *ode45* we give three initial values. The plot contains three curves corresponding to $y_i(t)$, $i = 1, 2, 3$. To get more readable graphs we have used the statement *text*.

The rhs-function *lrhs1.m* follows.

```
function out1 = lrhs1(t,y)
global k01 k21 k31
global k02 k12 k13
global I1
out1 = [-(k01 + k21 + k31)*y(1) + k12*y(2) + k13*y(3) + I1 ; ...
k21*y(1) - (k02 + k12)*y(2) ; k31*y(1)-k13*y(3)] ;
```

The assign statement for the output variable *out1* contains the three rhs from System (1.16) ordered and separated by ";".
For the first numerical experiment we have considered the initial conditions (1.17), $L = 600$ (days), $I_1 = 49.3\,\mu g/day$ and the following coefficients,

$$k_{01} = 0.0211, \quad k_{21} = 0.0111, \quad k_{31} = 0.0039,$$

$$k_{02} = 0.0162, \quad k_{12} = 0.0124, \quad k_{13} = 0.000035.$$

The graphs are given in Figure 1.20.

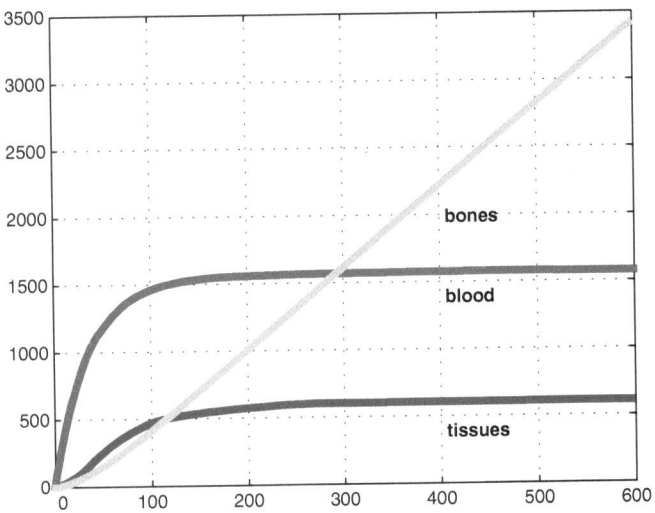

Fig. 1.20. The graphs of y_1, y_2, and y_3

For another experiment we replace the constant lead input rate I_1 by a time step function

$$I(t) = \begin{cases} I_1, & \text{for } t \in [0, t^*] \\ I_2, & \text{for } t \in (t^*, L], \end{cases} \tag{1.18}$$

where I_1 and I_2 are positive constants and $0 < t^* < L$. We introduce as global variables in the corresponding script file $I1$, $I2$, and tsw (for the switch time t^*). The rhs-function is

```
function out1 = lrhs2(t,y)
global k01 k21 k31
global k02 k12 k13
global I1 I2 tsw
if t > tsw
    a = I2 ;
else
```

```
        a = I1 ;
   end
   out1 = [−(k01 + k21 + k31)*y(1) + k12*y(2) + k13*y(3) + a ; ...
   k21*y(1) − (k02 + k12)*y(2) ; k31*y(1) − k13*y(3)] ;
```

We take $L = 800$, $t^* = 400$, $I_1 = 49.3$, and $I_2 = 2$. The other input values are as before and we get the graphs in Figure 1.21.

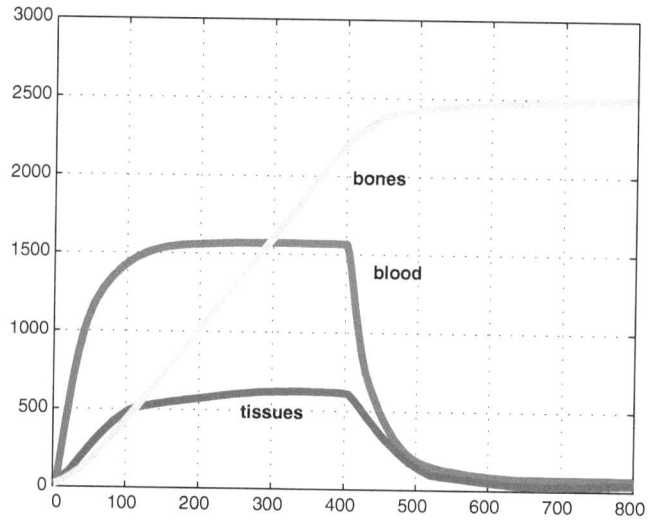

Fig. 1.21. The graphs of y_1, y_2, and y_3 for lead input given by (1.18)

We consider next a 4×4 model, that is, a model with four equations and four unknown functions.

An autocatalytic reaction

Such a model can be found, for instance, in [BC98, Section 5.1] and also in [http1]. Let us consider the species Y_i, $i = 1, 2, 3, 4$, and the autocatalytic reaction

$$Y_1 \xrightarrow{\ k_1\ } Y_2 \ ,$$
$$Y_2 \xrightarrow{\ k_2\ } Y_3 \ ,$$
$$Y_2 + 2Y_3 \xrightarrow{\ k_3\ } 3Y_3 \ ,$$
$$Y_3 \xrightarrow{\ k_4\ } Y_4 \ ,$$

with rate constants $k_i > 0$, $i = 1, 2, 3, 4$. Let $y_i(t)$ denote the concentration of Y_i at time $t \in [0, L]$. Then the model is given by the IVP:

$$\begin{cases} y_1' = -k_1 y_1, \\ y_2' = k_1 y_1 - k_2 y_2 - k_3 y_2 y_3^2, \\ y_3' = k_2 y_2 - k_4 y_3 + k_3 y_2 y_3^2, \\ y_4' = k_4 y_3, \\ y_1(0) = \alpha, \quad y_2(0) = y_3(0) = y_4(0) = 0, \end{cases} \tag{1.19}$$

for $t \in [0, L]$. Time and concentrations have been scaled to dimensionless form. The program is the following one.

```
% Autocatalytic reaction – auto1.m
% http://www.math.hmc.edu/resources/odes/odearchitect/examples/
clear
global k1 k2 k3 k4
load file4.txt
disp('get model parameters') ;
k1 = file4(1) ;
k2 = file4(2) ;
k3 = file4(3) ;
k4 = file4(4) ;
disp('get data') ;
L = input('final time : ') ;
alpha = input('y1(0) = ') ;
lw = input('LineWidth : ') ; % for graphical use (plot)
tspan = 0:0.01:L ;
[t y] = ode45('arerhs1',tspan,[alpha ; 0 ; 0 ; 0]) ;
z1 = y(:,1) / 200 ;
z4 = y(:,4) / 200 ;
plot(t,z1,'*',t,y(:,2),'r',t,y(:,3),'g',t,z4,'co') % plot 1
grid
xlabel('\bf t','FontSize',16)
text(50,2,'\bf y1/200')
text(100,0.25,'\bf y4/200')
text(250,2.5,'\bf y2')
text(200,0.2,'\bf y3')
figure(2) % plot 2
plot(t,y(:,2),'r','LineWidth',lw) ; grid
xlabel('\bf t','FontSize',16)
hold on
plot(t,y(:,3),'g','LineWidth',lw)
text(250,2.5,'\bf y2','FontSize',16)
text(200,0.2,'\bf y3','FontSize',16)
figure(3)
plot(y(:,2),y(:,3),'*') ; grid % plot 3
xlabel('\bf y2','FontSize',16)
ylabel('\bf y3','FontSize',16)
```

To get equivalent magnitudes for the graphical representation we have divided y_1 and y_4 by 200. We have also obtained the representations for $y_2(t)$ and $y_3(t)$ together and in the plane y_2Oy_3.
The rhs-function is

```
function out1 = arerhs1(t,y)
global k1 k2 k3 k4
out1 = [−k1*y(1) ; k1*y(1) − y(2)*(k2 + k3*y(3)∧2) ; ...
y(2)*(k2 + k3*y(3)∧2) − k4*y(3) ; k4*y(3)] ;
```

For the numerical tests we have considered $L = 400$, $\alpha = 500$, and rate constants to be $k_1 = 0.002$, $k_2 = 0.08$, and $k_3 = k_4 = 1$. The graphs are given as follows: all $y_i(t)$ with some scaling (*plot 1* from the program) in Figure 1.22; $y_2(t)$ and $y_3(t)$ (*plot 2* from the program) in Figure 1.23 (with $lw = 2$); the y_2Oy_3 graph (*plot 3* from the program) in Figure 1.24.
We now consider a model from mechanics. It is quite interesting from the point of view of numerical tests, more exactly in which concerns the adequacy of the model depending on the range of parameters.

A loaded beam model

We consider a beam that occupies the segment $[0, L]$ and is fixed at $x = 0$. At $x = L$ the beam is free and it is loaded by a weight P. We denote by $y(x)$ the displacement of the loaded beam for $x \in [0, L]$. The mathematical model is (see [DC72, Section 8.14]):

$$\begin{cases} \dfrac{y''}{[1 + (y')^2]^{3/2}} = \dfrac{P}{EI}(L - x), & x \in [0, L] \\ y(0) = y'(0) = 0. \end{cases} \qquad (1.20)$$

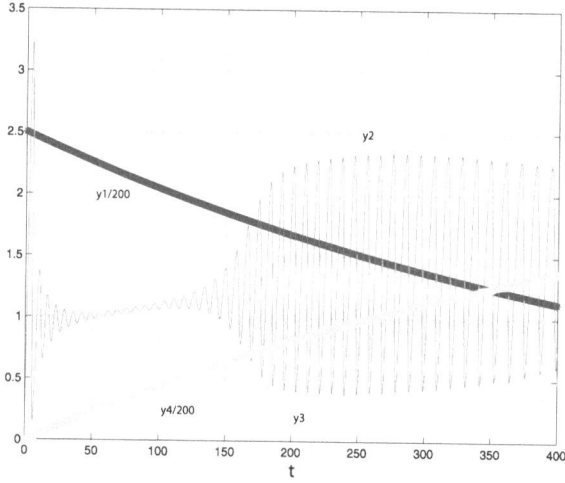

Fig. 1.22. The graphs of $y_1/200$, y_2, y_3, and $y_4/200$

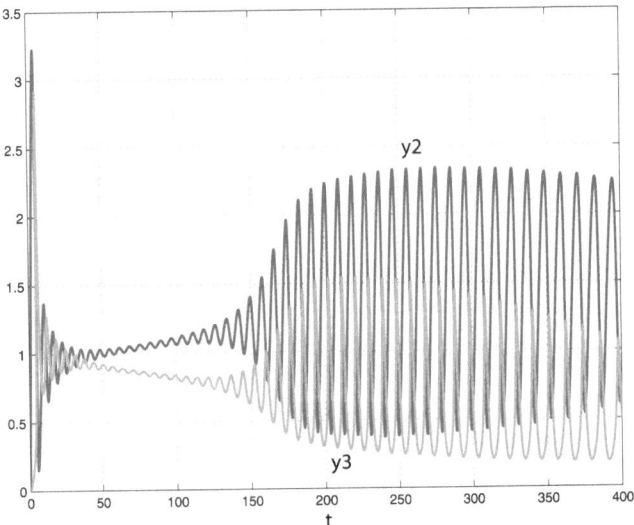

Fig. 1.23. The graphs of y_i, $i \in \{2, 3\}$

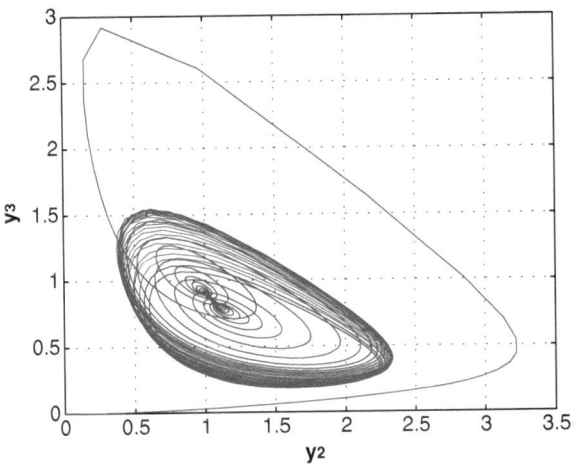

Fig. 1.24. Representation of y_i, $i \in \{2, 3\}$ in (y_2, y_3) coordinates

Let us point out that $y(x)$ represents the absolute value of the displacement and therefore we have to consider $-y(x)$ for the *plot* statement because the displacement is made in the negative sense of the Oy axis. In (1.20) E denotes the Young's modulus and I the moment of inertia of the cross-section of the beam. Because we consider these physical values to be constant (independent of x), we simply denote $C = P/(EI)$ in the sequel.

We now consider two cases.

(1) The weight P is "small" and therefore the velocity of the displacement is "small." Hence $1 + (y')^2 \approx 1$ and (1.20) is replaced by

$$\begin{cases} y'' = C(L - x), & x \in [0, L] \\ y(0) = y'(0) = 0. \end{cases} \tag{1.21}$$

Integrating we get the solution

$$y(x) = \frac{C}{6}x^2(3L - x), \quad x \in [0, L]. \tag{1.22}$$

(2) The weight P is not "small" and therefore $(y')^2$ cannot be neglected. We have to integrate Problem (1.20) numerically. We introduce as usual the functions $y_1 = y$, $y_2 = y'$, and we get the system

$$\begin{cases} y_1' = y_2, & x \in [0, L] \\ y_2' = C(1 + y_2^2)^{3/2} \cdot (L - x), & x \in [0, L] \\ y_1(0) = y_2(0) = 0. \end{cases} \tag{1.23}$$

The script file graphically compares the solution (1.22) for the simplified model (1.21) to the numerical solution obtained with *ode45* for Problem (1.23). Let us remark that the polynomial in Formula (1.22) may always be computed for an interval $[0, L]$.

```
% file beam1.m
% the loaded beam fixed at one end
% the numerical integration is made by ode45
% a comparison is made to the simplified model
clear
global L C
L = input('L : ')
C = input('C : ')
L3 = 3.0 * L ;
C6 = C/6.0 ;
z = 0:0.001:L ;
[x y] = ode45('b', z, [0 ; 0]) ;
plot(x,-y(:,1),'s') ; grid
xlabel('\bf x','FontSize',16)
ylabel('\bf -y(x)','FontSize',16)
hold on
w = C6*(z.^2).*(L3 - z) ;
plot(z,-w,'r*')
legend('numerical','polynomial',0)
hold off
```

Here $[x\ y]$ is the pair of vectors returned by *ode45* and $[z\ w]$ is the pair of vectors corresponding to the simplified model (Formula (1.22)). Remark that

w is computed from z by array-smart operations because z is a vector. The rhs for *ode45* is given by the function file *b.m* taking into account the right-hand side for the equations of system (1.23).

```
function ret = b(x,y)
% the rhs for beam1.m
global L C
ret = [y(2) ; C*(L − x)*(1.0 + y(2)∧2)∧1.5] ;
```

To compare the values obtained at $x = L$ we add the following sequence to the script file *beam1.m*.

```
format long
N = length(x) ;
ynum = y(N,1) % numerical value at x = L
M = length(w) ;
ypol = w(M) % polynomial value at x = L
```

We have made numerical tests for $L = 1$ and for different values of the constant C. Let us recall that $C = P/(EI)$ and therefore in our numerical experiments C has the same meaning as the weight P. For $C = 1$ we get Figure 1.25. There is a difference between the two solutions when approaching the right limit of the interval. We have used the statement *text* to insert alphanumerical strings inside the figure. Here "numer" stands for "numerical solution" and "polyn" stands for "polynomial solution".

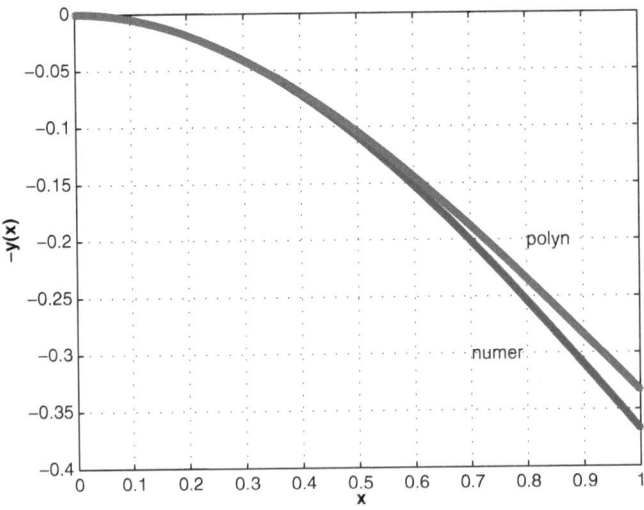

Fig. 1.25. Numerical and polynomial solutions for $C = 1$

For $C = 1.75$ we get Figure 1.26. The difference between the two solutions is more important. The graph of the numerical solution obtained by *ode45* is

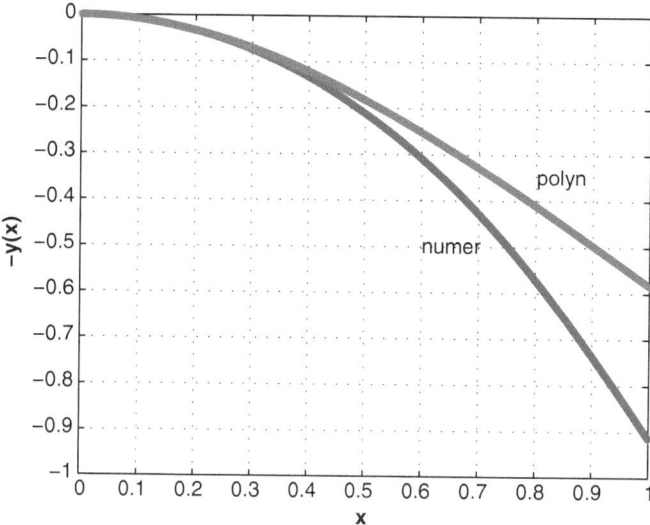

Fig. 1.26. Numerical and polynomial solutions for $C = 1.75$

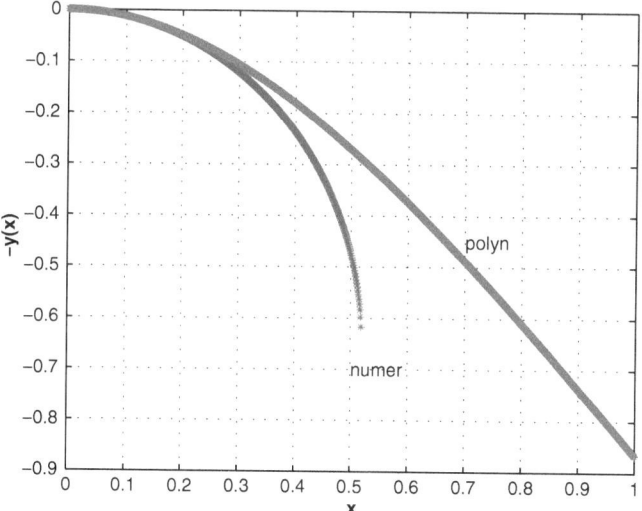

Fig. 1.27. Numerical and polynomial solutions for $C = 2.6$

below the graph of the polynomial solution. As expected the simplified model does not work properly for C large.

The last experiment was made for $C = 2.6$ and the result may be seen in Figure 1.27. The numerical solution failed for some time value which means from a physical point of view that the beam was broken due to the heavy load. The polynomial solution was obtained because the polynomial (1.22) is defined for any x, but it is meaningless. It is quite clear that the polynomial model is not adequate for large values of the parameter C (force).

1.8 3D Graphics

The 3D equivalent of *plot* is *plot3*. It is useful for functions defined by a parametric formula. The statement

plot3(x,y,z)

is valid if x, y, z are vectors of equal length; denote it by N. The statement *plot3* produces the 3D curve passing through the points (x_i, y_i, z_i), $i = 1, 2, \ldots, N$. We cite here a nice simple example that can be found in the MATLAB documentation, namely the helix defined by the parametric equations

$$\begin{cases} x = \sin t \\ y = \cos t \\ z = t \end{cases} \tag{1.24}$$

for $t \in [0, 6\pi]$. The script file is *helix.m*

```
% file helix.m
% parametric representation with plot3
h = pi/1000 ;
t = 0:h:6*pi ;
plot3(sin(t),cos(t),t,'*') ; grid
xlabel('\bf sin(t)','FontSize',16)
ylabel('\bf cos(t)','FontSize',16)
zlabel('\bf t','FontSize',16)
```

and the result is shown in Figure 1.28.

If we now turn back to the autocatalytic reaction from Section 1.7 (program *auto1.m*) and we add the sequence

```
figure(4)
plot3(y(:,2),y(:,3),t)
```

we get the Figure 1.29 (the model we have considered deals with oscillations in a wineglass).

To plot the graph of a function $z = f(x, y)$ the corresponding statement is *mesh* or *surf*. Suppose that f is defined on $[a, b] \times [c, d]$. We first have to build the vectors x and y that contain the grid points corresponding, respectively, to Ox and Oy axes.

```
x = a:hx:b ;
y = c:hy:d ;
```

Then we make the matrices X and Y by

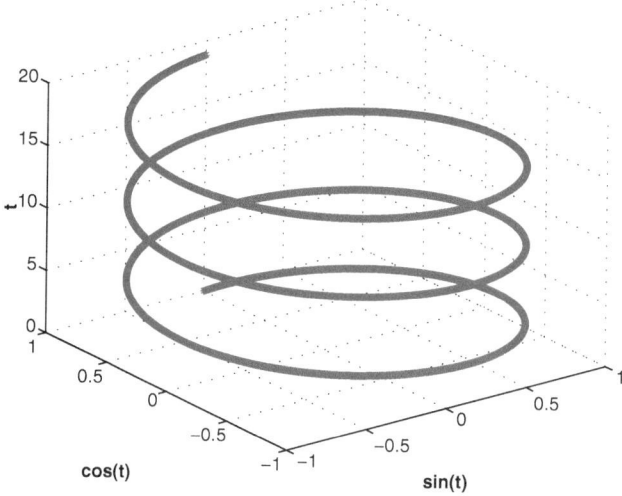

Fig. 1.28. The parametric curve given by (1.24)

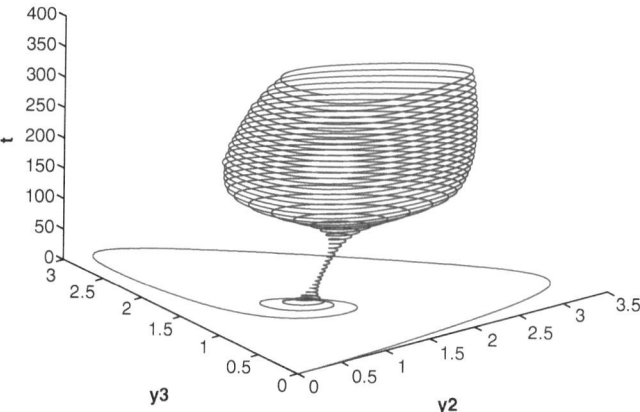

Fig. 1.29. Oscillations in a wineglass

[X,Y] = meshgrid(x,y) ;

Assume that $length(x)$ is n and $length(y)$ is m. Then both matrices are $m \times n$. All m rows of X are equal to vector x and all columns of Y are equal to vector y. Then we make the matrix Z by

Z = f(X,Y) ;

where $f.m$ is the corresponding array-smart function. Therefore a 3D "wire mesh surface" is generated by the statement

mesh(X,Y,Z)

and a 3D "faceted surface" is generated by the statement

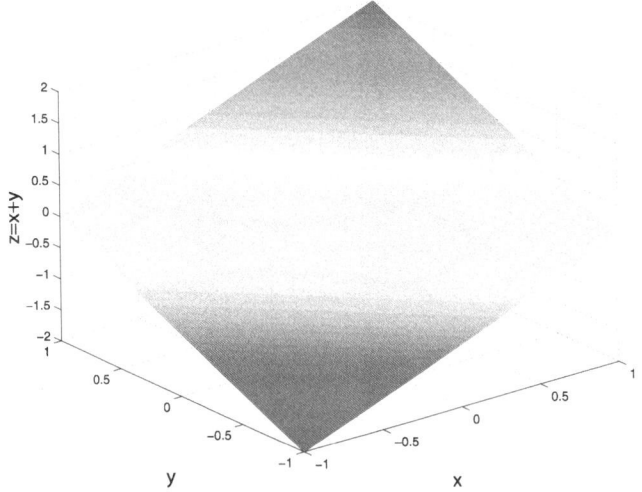

Fig. 1.30. The plane of equation $z = x + y$

    surf(X,Y,Z)

It is also possible to write directly

    mesh(X,Y,f(X,Y))

and the same for *surf*.

As a first example we generate a plane (Figure 1.30). Here is the script file

```
% file graf1.m ; 3D representation of a plan
clear
x = −1:0.01:1 ;
y = x ;
[X,Y] = meshgrid(x,y) ;
mesh(X,Y,X + Y)
xlabel('\bf x','FontSize',16)
ylabel('\bf y','FontSize',16)
zlabel('\bf z=x+y','FontSize',16)
```

Let us now consider the function $f(x, y) = x^2 + y^2$, $(x, y) \in [-3, 3] \times [-3, 3]$. The script file is *graf4.m*.

```
% file graf4.m ; graph of the function z = x*x + y*y
clear
h = 0.1
x = −3:h:3 ;
y = x ;
[X,Y] = meshgrid(x,y) ;
```

```
Z = X.^2 + Y.^2 ;
mesh(X,Y,Z) ; title('\bf MESH','FontSize',16)
xlabel('\bf x','FontSize',16)
ylabel('\bf y','FontSize',16)
zlabel('\bf z(x,y)','FontSize',16)
figure(2)
surf(X,Y,Z) ; title('\bf SURF','FontSize',16)
xlabel('\bf x','FontSize',16)
ylabel('\bf y','FontSize',16)
zlabel('\bf z(x,y)','FontSize',16)
```

The program above produces two figures: figure (1) is implicit and contains the result of the *mesh* statement, and figure (2) is declared and contains the result of the *surf* statement (see Figures 1.31 and 1.32).

Bibliographical Notes and Remarks

Several mathematical models described by ODEs have been investigated in the literature. We refer for instance to [CLW69], [DC72], [GW84], [PFT90], and [Ist05]. For biological models see Section 1.7 and references therein ([BC98], [Mur89], [Smo83], and [http1]). For applications together with MATLAB programming we cite [Coo98] and [QS03]. Many very nice examples of such programs can also be found in the MathWorks documentation.

For numerical linear algebra we recommend [You71] and [Cia94]. Numerical methods for ODEs may be found, for instance, in [CM89]. Existence and

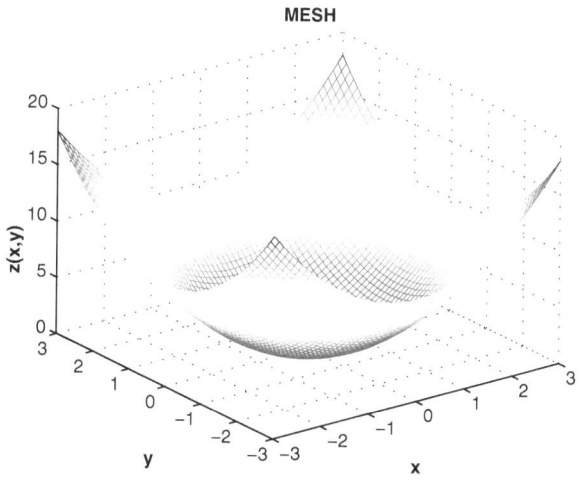

Fig. 1.31. Figure (1)

SURF

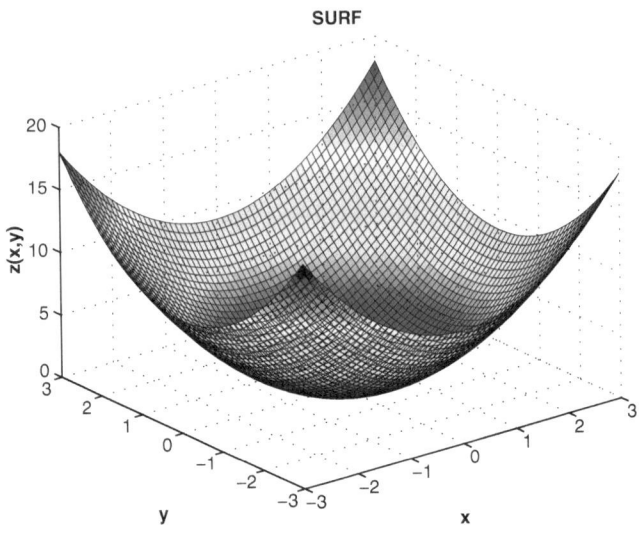

Fig. 1.32. Figure (2)

uniqueness results for IVPs governed by ODEs (Theorem 1.1) can be found
in any classical book on differential equations (e.g., [Har02] or [Zwi97]).

For examples of 3D graphics and other examples we refer to [Coo98].

Exercises

1.1. Consider the IVP:

$$\begin{cases} y'(x) = x + y(x), \\ y(0) = 1, \end{cases}$$

which has the mathematical solution

$$y(x) = 2e^x - (x + 1).$$

Solve it numerically on some given interval $[0, L]$ and make a figure that
contains the graphs of the mathematical solution and of the numerical one.
Compare, with *format long*, $y(L)$ with the numerical value obtained for $x = L$.
Do it for different increasing values of L.

1.2. The same problem as above for the IVP:

$$\begin{cases} y'(x) = -y(x), \\ y(0) = 1, \end{cases}$$

which has the mathematical solution

$$y(x) = e^{-x}.$$

1.3. Consider the IVP:

$$\begin{cases} y'(x) = x^2 + y(x)^2, \\ y(0) = -1. \end{cases}$$

For some interval $[0, L]$, compare the numerical solutions obtained with *ode23* and *ode45*. Plot the corresponding graphs on the same figure and compare the numerical values obtained for $x = L$. Do it for different increasing values of L.

1.4. Consider the following IVP, which gives the evolution of an insect population (such as the spruce budworm model) subject to a harvesting process:

$$\begin{cases} N'(t) = rN(t) \left(1 - \dfrac{N(t)}{K}\right) - u(t)N(t), \quad t \in [0, T] \\ N(0) = N_0. \end{cases}$$

Take, as for the first test of the spruce budworm model, $r = 0.3$, $K = 5$, $T = 20$, and $N_0 = 30$. Make the graph of the insect population $N(t)$ for $u(t) = t$ on $[0, T]$. Also compute the harvest given by $\int_0^T u(t)N(t)dt$.

Hint. Let the rhs-function be *sbw3*. Hence

```
function z = sbw3(x,y)
global r K
z = r*y*(1 - y/K) - x*y ;
```

To compute the harvest value we use the statement

```
[x y] = ode45('sbw3', [0 T], N0) ;
...
harv = trapz(x,x.*y)
```

1.5. Plot the surface of equation $z = \sin(x - y)$, $x \in [0, 4]$, $y \in [0, 6]$.

Hint. Use the script file *graf2.m*.

```
% file graf2.m
% 3D graph for z = sin(x - y)
clear
h = 0.01
x = 0:h:4 ;
y = 0:h:6 ;
[X,Y] = meshgrid(x,y) ;
Z = sin(X - Y) ;
mesh(X,Y,Z)
xlabel('\bf x','FontSize',16)
ylabel('\bf y','FontSize',16)
zlabel('\bf z = sin(x - y)','FontSize',16)
```

1.6. Plot the graph of the function f defined by

$$f(x,y) = e^{-((x-3)^2+(y-2)^2)}, \quad x \in [0,4], \quad y \in [0,6].$$

Hint. Use the script file *graf3.m.*

```
% file graf3.m
% uses the function f2D
% 3D graphic for z = f2D(x,y)
clear
h = 0.01
x = 0:h:4 ;
y = 0:h:6 ;
[X,Y] = meshgrid(x,y) ;
Z = f2D (X,Y) ;
mesh(X,Y,Z)
xlabel('\bf x','FontSize',16)
ylabel('\bf y','FontSize',16)
zlabel('\bf z = f2D(x,y)','FontSize',16)
```

and the array-smart function *f2D.m* which follows.

```
function z = f2D(x,y)
q = (x - 3).^2 + (y - 2).^2 ;
z = exp(-q) ;
```

2

Optimal control of ordinary differential systems. Optimality conditions

This chapter and the next one are devoted to some basic ideas and techniques in optimal control theory of ordinary differential systems. We do not treat the optimal control problem or Pontryagin's principle in their most general form; instead we prefer a direct approach for some significant optimal control problems in life sciences and economics governed by ordinary differential systems. We point out the main steps in the study of an optimal control problem for each investigated example. These steps are similar for all examples. There are, however, specific technical difficulties for each investigated problem.

The main goal of this chapter is to prove the existence of an optimal control and to obtain first-order necessary conditions of optimality (Pontryagin's principle) for some significant optimal control problems. The necessary optimality conditions give valuable information about the structure of the optimal control. Numerical algorithms to approximate the optimal control and corresponding MATLAB® programs are indicated.

A general formulation of Pontryagin's principle for optimal control problems related to ordinary differential systems can be found in [Bar93] and [Bar94].

2.1 Basic problem. Pontryagin's principle

A quite general optimal control problem governed by an ordinary differential system can be formulated in the following form,

$$\text{Maximize } \mathcal{L}(u, x^u) = \int_0^T G(t, u(t), x^u(t))dt + \varphi(x^u(T)), \qquad \textbf{(P1)}$$

subject to $u \in K \subset L^2(0, T; \mathbb{R}^m)$ $(T > 0)$, where x^u is the Carathéodory solution to

$$\begin{cases} x'(t) = f(t, u(t), x(t)), & t \in (0, T) \\ x(0) = x_0. \end{cases} \qquad (2.1)$$

S. Aniţa et al. *An Introduction to Optimal Control Problems in Life Sciences and Economics*, Modeling and Simulation in Science, Engineering and Technology, DOI 10.1007/978-0-8176-8098-5_2, © Springer Science+Business Media, LLC 2011

Here

$$G : [0,T] \times \mathbb{R}^m \times \mathbb{R}^N \to \mathbb{R},$$
$$\varphi : \mathbb{R}^N \to \mathbb{R},$$
$$f : [0,T] \times \mathbb{R}^m \times \mathbb{R}^N \to \mathbb{R}^N,$$

$x_0 \in \mathbb{R}^N$, $m, N \in \mathbb{N}^*$, and $K \subset L^2(0,T;\mathbb{R}^m)$ is a closed convex subset. From now all elements of an $\mathbb{R}^n$, $n \in \mathbb{N}^*$, are considered as column vectors.
Recall that a Carathéodory solution (we call it simply a solution) to (2.1) is a function x^u that belongs to $AC([0,T];\mathbb{R}^N)$ (see Appendix A.3), and satisfies

$$\begin{cases} (x^u)'(t) = f(t, u(t), x^u(t)) \text{ a.e. } t \in (0,T) \\ x^u(0) = x_0. \end{cases}$$

$L^2(0,T;\mathbb{R}^m)$ is the set of the controllers.

   This optimal control problem can be reformulated as the following minimization problem,

$$\text{Minimize } \{-\mathcal{L}(u, x^u)\},$$

subject to $u \in K \subset L^2(0,T;\mathbb{R}^m)$.

We assume here that for any $u \in L^2(0,T;\mathbb{R}^m)$, Problem (2.1) admits a unique solution, denoted by x^u. Equation (2.1) is called the state problem (equation).

• $u (\in K)$ is called the control (or controller). This is a constrained control because $u \in K$, and K is a subset of $L^2(0,T;\mathbb{R}^m)$.
• x^u is the state corresponding to the control u, and the mapping.
• $u \mapsto \mathcal{L}(u, x^u) = \Phi(u)$ is the cost functional.

We say that $u^* \in K$ is an optimal control for Problem (P1) if

$$\mathcal{L}(u^*, x^{u^*}) \geq \mathcal{L}(u, x^u),$$

for any $u \in K$. The pair (u^*, x^{u^*}) is called an optimal pair and $\mathcal{L}(u^*, x^{u^*})$ is the optimal value of the cost functional. We also say that (u^*, x^*) is an optimal pair if u^* is an optimal control and $x^* = x^{u^*}$.

Let $u^* \in K$ be an optimal control for (P1); that is,

$$\int_0^T G(t, u^*(t), x^{u^*}(t))dt + \varphi(x^{u^*}(T)) \geq \int_0^T G(t, u(t), x^u(t))dt + \varphi(x^u(T)),$$

for any $u \in K$.

We assume that the following succession of operations and arguments is allowed (under certain hypotheses – including Gâteaux differentiability (see Appendix A.1.5) – on G, φ, and f). We use the notations:

$$\begin{cases} f_u = \dfrac{\partial f}{\partial u}, & f_x = \dfrac{\partial f}{\partial x} \\[2mm] G_u = \dfrac{\partial G}{\partial u}, & G_x = \dfrac{\partial G}{\partial x} \\[2mm] \varphi_x = \dfrac{\partial \varphi}{\partial x} \end{cases}$$

(see also Appendix A.1.5). Here G_u, G_x, and φ_x are considered as column vectors.

Assume that the function defined on $L^2(0,T;\mathbb{R}^m)$, $u \mapsto x^u$ is everywhere Gâteaux differentiable. We denote this differential by dx^u.

Consider

$$V = \{v \in L^2(0,T;\mathbb{R}^m); \ u^* + \varepsilon v \in K \text{ for any } \varepsilon > 0 \text{ sufficiently small}\}.$$

For any $v \in V$, we define $z = dx^{u^*}(v)$; z is the solution to

$$\begin{cases} z'(t) = f_u(t,u^*(t),x^{u^*}(t))v(t) + f_x(t,u^*(t),x^{u^*}(t))z(t), & t \in (0,T) \\ z(0) = 0. \end{cases} \quad (2.2)$$

For an arbitrary but fixed $v \in V$ we have that

$$\int_0^T G(t,u^*(t),x^{u^*}(t))dt + \varphi(x^{u^*}(T)) \geq \int_0^T G(t,u^*(t) + \varepsilon v(t), x^{u^*+\varepsilon v}(t))dt$$
$$+ \varphi(x^{u^*+\varepsilon v}(T)),$$

and consequently

$$\int_0^T \frac{1}{\varepsilon}\left[G(t,u^*(t)+\varepsilon v(t), x^{u^*+\varepsilon v}(t)) - G(t,u^*(t),x^{u^*}(t)) \right] dt$$
$$+ \frac{1}{\varepsilon}\left[\varphi(x^{u^*+\varepsilon v}(T)) - \varphi(x^{u^*}(T)) \right] \leq 0,$$

for any $v \in V$, and for any $\varepsilon > 0$ sufficiently small.
We pass to the limit in the last inequality ($\varepsilon \to 0+$) and we get that

$$\int_0^T [v(t) \cdot G_u(t,u^*(t),x^{u^*}(t)) + z(t) \cdot G_x(t,u^*(t),x^{u^*}(t))]dt$$
$$+ z(T) \cdot \varphi_x(x^{u^*}(T)) \leq 0 \quad (2.3)$$

(here $\cdot$ denotes the usual scalar product on $\mathbb{R}^m$ as well as on $\mathbb{R}^N$), for any $v \in V$.

Let p be the Carathéodory solution (we assume that this solution exists and is unique), that we simply call the solution, to the adjoint problem (equation):

$$\begin{cases} p'(t) = -f_x^*(t,u^*(t),x^{u^*}(t))p(t) - G_x(t,u^*(t),x^{u^*}(t)), & t \in (0,T) \\ p(T) = \varphi_x(x^{u^*}(T)) \end{cases} \quad (2.4)$$

(p is called the adjoint state; the equation in (2.4) is linear).
Recall that if $A : {\rm I\!R}^k \longrightarrow {\rm I\!R}^s$ is a linear (and bounded) operator (A may be identified with a matrix, also denoted by A), then its adjoint operator A^* : ${\rm I\!R}^s \longrightarrow {\rm I\!R}^k$ (also linear and bounded) may be identified with the transpose of matrix A, and denoted also by A^* (or A^T).

By multiplying (2.2) by p and integrating by parts on $[0,T]$ we get that

$$z(T) \cdot p(T) - \int_0^T z(t) \cdot p'(t)dt$$
$$= \int_0^T [f_u(t, u^*(t), x^{u^*}(t))v(t) + f_x(t, u^*(t), x^{u^*}(t))z(t)] \cdot p(t)dt,$$

for any $v \in V$. By (2.4) we obtain that

$$z(T) \cdot \varphi_x(x^{u^*}(T))$$
$$+ \int_0^T z(t) \cdot [f_x^*(t, u^*(t), x^{u^*}(t))p(t) + G_x(t, u^*(t), x^{u^*}(t))]dt$$
$$= \int_0^T [v(t) \cdot f_u^*(t, u^*(t), x^{u^*}(t))p(t) + z(t) \cdot f_x^*(t, u^*(t), x^{u^*}(t))p(t)]dt,$$

and consequently

$$\int_0^T z(t) \cdot G_x(t, u^*(t), x^{u^*}(t))dt + z(T) \cdot \varphi_x(x^{u^*}(T))$$
$$= \int_0^T v(t) \cdot f_u^*(t, u^*(t), x^{u^*}(t))p(t)dt,$$

for any $v \in V$. By (2.3) we finally get that

$$\int_0^T v(t) \cdot [G_u(t, u^*(t), x^{u^*}(t)) + f_u^*(t, u^*(t), x^{u^*}(t))p(t)]dt \le 0,$$

for any $v \in V$, which means

$$G_u(\cdot, u^*, x^{u^*}) + f_u^*(\cdot, u^*, x^{u^*})p \in N_K(u^*), \qquad (2.5)$$

where $N_K(u^*)$ is the normal cone at K in u^* (see Appendix A.1.4).
We get the same conclusion if we multiply (2.4) by z (after a similar argumentation).

Equations (2.1), (2.4), and (2.5) represent Pontryagin's (or maximum) principle and (2.4) and (2.5) are the first-order necessary conditions of optimality (optimality conditions) for the given optimal control problem.

The main goal now is to use the maximum principle in order to calculate an optimal control u^* or to approximate it by using an appropriate numerical scheme. In order to use Condition (2.5) we need to determine the set $N_K(u^*)$.

If we take, for example, $K = L^2(0,T; I\!R^m)$, then for any $u \in K = L^2(0,T; I\!R^m)$, $N_K(u) = \{0\} \subset L^2(0,T; I\!R^m)$.

If we take $m = 1$, and

$$K = \{w \in L^2(0,T); \ L_1 \leq w(t) \leq L_2 \text{ a.e. } t \in (0,T)\},$$

where $L_1, L_2 \in I\!R$, $L_1 < L_2$, then for any $u \in K$ we have

$$N_K(u) = \{w \in L^2(0,T); \quad w(t) \geq 0 \ \text{ if } u(t) = L_2, w(t) \leq 0 \text{ if } u(t) = L_1,$$
$$w(t) = 0 \ \text{ if } L_1 < u(t) < L_2 \text{ a.e. } t \in (0,T)\}$$

(see Appendix A.1.4).

A general scheme to prove the existence of an optimal control u^* is the following one.
Let

$$d = \sup_{u \in K} \mathcal{L}(u, x^u) \in I\!R.$$

For any $n \in I\!N^*$, there exists $u_n \in K$, such that

$$d - \frac{1}{n} < \mathcal{L}(u_n, x^{u_n}) \leq d.$$

Step 1: Prove that there exists a subsequence $\{u_{n_k}\}$ such that

$$u_{n_k} \longrightarrow u^* \text{ weakly in } L^2(0,T; I\!R^m).$$

If for example, K is bounded, then the last conclusion follows immediately. Inasmuch as K is a closed convex subset of $L^2(0,T; I\!R^m)$, K is also weakly closed, and consequently $u^* \in K$.

Step 2: Prove that there exists a subsequence of $\{x^{u_{n_k}}\}$, denoted by $\{x^{u_{n_r}}\}$, convergent to x^{u^*} in $C([0,T]; I\!R^N)$ (sometimes the convergence in $L^2(0,T; I\!R^N)$ is enough).

Step 3: From

$$d - \frac{1}{n_r} < \mathcal{L}(u_{n_r}, x^{u_{n_r}}) \leq d,$$

we get (by passing to the limit) that

$$\mathcal{L}(u^*, x^{u^*}) = d,$$

and consequently u^* is an optimal control for problem (P1).
Notice that we can derive (2.4) and (2.5) by using the Hamiltonian H, defined by

$$H(t, u, x, p) = G(t, u, x) + f(t, u, x) \cdot p.$$

If we take

$$x' = H_p$$

we get the state equation. By

$$p' = -H_x$$

we get the adjoint equation and by

$$H_u \in N_K(u^*),$$

we get (2.5).

Let us mention that some authors consider the following problem as the adjoint problem:

$$\begin{cases} p'(t) = -f_x^*(t, u^*(t), x^{u^*}(t))p(t) + G_x(t, u^*(t), x^{u^*}(t)), & t \in (0, T) \\ p(T) = -\varphi_x(x^{u^*}(T)) \end{cases}$$

The solution to this problem is $p = -\tilde{p}$, where $\tilde{p}$ is the solution to (2.4).

Condition (2.5) becomes

$$G_u(\cdot, u^*, x^{u^*}) - f_u^*(\cdot, u^*, x^{u^*})p \in N_K(u^*),$$

and the Hamiltonian H is:

$$H(t, u, x, p) = -G(t, u, x) + f(t, u, x) \cdot p.$$

We, however, use both conventions (for the adjoint problem) in the next chapters.

In most situations (2.1) appears as a semilinear problem; that is, f has the following form,

$$f(t, u, x) = Ax + \tilde{f}(t, u, x),$$

where $A : \mathbb{R}^N \longrightarrow \mathbb{R}^N$ is a (particular) linear operator. Then

$$f_u = \tilde{f}_u, \quad f_x = A + \tilde{f}_x.$$

Several optimal control problems related to age-structured models, semilinear parabolic equations or to integroparabolic equations may be written in the abstract form (P1)–(2.1), where

$$G : [0, T] \times U \times X \to \mathbb{R},$$
$$\varphi : X \to \mathbb{R}$$

$(U, X$ are appropriate real Hilbert spaces), $x_0 \in X$, and $K \subset L^2(0, T; U)$ is a closed convex subset. Here f has the above mentioned form, and A is a linear (possibly unbounded) operator, $A : D(A) \subset X \rightarrow X$ (see Chapters 4 and 5).

Here we have presented only a general scheme and not a rigorous proof of the maximum principle.

In the next sections we illustrate how this scheme works for significant examples of optimal control problems in life sciences and economics governed by ordinary differential systems. We deduce the maximum principle again for all these examples in a rigorous manner. We use the maximum principle to calculate or to approximate optimal control. Chapters 4 and 5 are devoted to control problems governed by partial differential equations. As announced the scheme is the same, but there are, of course, more technical difficulties.

2.2 Maximizing total consumption

We consider a mathematical model of a simplified economy. Let $x(t)$ be the rate of production at the moment $t \geq 0$ (the economical output). We have

$$x(t) = I(t) + C(t), \quad t \geq 0,$$

where

- $I(t)$ is the rate of investment at the moment t.
- $C(t)$ is the rate of consumption at the moment t.

Denote by $u(t) \in [0, 1]$ the part of production $x(t)$ that is allocated to investment at moment t; that is,

$$I(t) = u(t)x(t).$$

We obtain that

$$C(t) = (1 - u(t))x(t), \quad t \geq 0.$$

We deal with the simple case when the production growth rate is proportional to the rate of investment. This means

$$x'(t) = \gamma u(t)x(t),$$

where $\gamma \in (0, +\infty)$.

We introduce a "utility" function $F(C)$, and we wish to find out the control that maximizes the welfare integral

$$\int_0^T e^{-\delta t} F(C(t))dt.$$

Here $T > 0$, and $\delta \geq 0$ is a discount rate (a measure of preference for earlier rather than later consumption).

We simplify our model by taking $F(C) = C$ and $\delta = 0$. The total consumption on the time interval $[0, T]$ is

$$\int_0^T C(t)dt = \int_0^T (1 - u(t))x(t)dt.$$

We therefore obtain the following optimal control problem (see [Bar94]),

$$\text{Maximize } \int_0^T (1 - u(t))x^u(t)dt, \qquad \text{(P2)}$$

subject to $u \in L^2(0, T)$, $0 \leq u(t) \leq 1$ a.e. $t \in (0, T)$, where x^u is the solution of

$$\begin{cases} x'(t) = \gamma u(t)x(t), & t \in (0, T) \\ x(0) = x_0 > 0. \end{cases} \qquad (2.6)$$

The problem seeks to find the control u that maximizes total consumption on the time interval $[0, T]$.

The solution x^u to (2.6) is given by

$$x^u(t) = x_0 \exp\left(\int_0^t \gamma u(s)ds\right), \quad t \in [0, T].$$

Problem (P2) is a particular case of (P1), for $m = 1$, $N = 1$,

$$G(t, u, x) = (1 - u)x,$$

$$\varphi(x) = 0,$$

$$f(t, u, x) = \gamma ux$$

and

$$K = \{w \in L^2(0, T); \ 0 \leq w(t) \leq 1 \text{ a.e. } t \in (0, T)\}.$$

Existence of an optimal pair for (P2)

Define

$$\Phi(u) = \int_0^T (1 - u(t))x^u(t)dt, \quad u \in K$$

and let

$$d = \sup_{u \in K} \Phi(u).$$

Because for any $u \in K$ we have that

$$0 < x^u(t) \le x_0 e^{\gamma t}, \quad t \in [0, T],$$

then we get that

$$0 \le \Phi(u) = \int_0^T (1 - u(t)) x^u(t) dt \le x_0 T e^{\gamma T}.$$

In conclusion $d \in [0, +\infty)$.

So, for any $n \in \mathbb{N}^*$, there exists $u_n \in K$ such that

$$d - \frac{1}{n} < \Phi(u_n) \le d. \tag{2.7}$$

K is a bounded subset of $L^2(0, T)$, therefore it follows that there exists a subsequence $\{u_{n_k}\}_{k \in \mathbb{N}^*}$ such that

$$u_{n_k} \longrightarrow u^* \quad \text{weakly in } L^2(0, T). \tag{2.8}$$

The limit u^* belongs to K because K is a closed convex subset of $L^2(0, T)$, and so it is weakly closed. The last convergence and the explicit formula for x^u imply that

$$x^{u_{n_k}} \longrightarrow x^{u^*} \quad \text{in } L^2(0, T). \tag{2.9}$$

By (2.7) we get that

$$d - \frac{1}{n_k} < \int_0^T (1 - u_{n_k}(t)) x^{u_{n_k}}(t) dt \le d \quad \text{for any } k \in \mathbb{N}^*. \tag{2.10}$$

By (2.8) and (2.9) we obtain (we pass to the limit in (2.10)) that

$$d = \int_0^T (1 - u^*(t)) x^{u^*}(t) dt,$$

that is, (u^*, x^{u^*}) is an optimal pair (and u^* is an optimal control) for (P2).

In order to simplify the notations we denote $x^* := x^{u^*}$.

The maximum principle

For an arbitrary but fixed $v \in V = \{w \in L^2(0, T);\ u^* + \varepsilon w \in K$ for any $\varepsilon > 0$ sufficiently small$\}$ we denote by z the solution to

$$\begin{cases} z'(t) = \gamma u^*(t) z(t) + \gamma v(t) x^*(t), & t \in (0, T) \\ z(0) = 0. \end{cases} \tag{2.11}$$

z is given by

$$z(t) = \int_0^t \exp\{\int_s^t \gamma u^*(\tau)d\tau\}\gamma v(s)x^*(s)ds, \quad t \in [0,T]. \qquad (2.12)$$

Inasmuch as

$$\int_0^T (1 - u^*(t))x^*(t)dt \geq \int_0^T (1 - u^*(t) - \varepsilon v(t))x^{u^*+\varepsilon v}(t)dt,$$

for any $\varepsilon > 0$ sufficiently small, we get that

$$\int_0^T [(1 - u^*(t))\frac{x^{u^*+\varepsilon v}(t) - x^*(t)}{\varepsilon} - v(t)x^{u^*+\varepsilon v}(t)]dt \leq 0. \qquad (2.13)$$

Let us prove that

$$x^{u^*+\varepsilon v} \longrightarrow x^* \quad \text{in } C([0,T])$$

and

$$\frac{x^{u^*+\varepsilon v} - x^*}{\varepsilon} \longrightarrow z \quad \text{in } C([0,T]),$$

as $\varepsilon \to 0+$.

Indeed, for any $\varepsilon > 0$ sufficiently small we have

$$x^{u^*+\varepsilon v}(t) = x_0 \exp\{\gamma \int_0^t (u^*(s) + \varepsilon v(s))ds\}$$

$$= x^{u^*}(t) \exp\{\varepsilon\gamma \int_0^t v(s)ds\}, \quad t \in [0,T],$$

which implies that

$$|x^{u^*+\varepsilon v}(t) - x^{u^*}(t)| = |x^{u^*}(t)| \cdot |\exp\{\varepsilon\gamma \int_0^t v(s)ds\} - 1|, \quad t \in [0,T].$$

Because

$$|\exp\{\varepsilon\gamma \int_0^t v(s)ds\} - 1| \longrightarrow 0,$$

uniformly on $[0,T]$, we may infer that

$$x^{u^*+\varepsilon v} \longrightarrow x^* \quad \text{in } C([0,T]).$$

For any $\varepsilon > 0$ sufficiently small we consider

$$w_\varepsilon(t) = \frac{x^{u^*+\varepsilon v} - x^*}{\varepsilon} - z(t), \quad t \in [0,T].$$

w_ε is the solution to

$$\begin{cases} w'(t) = \gamma u^*(t)w(t) + \gamma v(t)[x^{u^*+\varepsilon v}(t) - x^{u^*}(t)], & t \in (0,T) \\ w(0) = 0, \end{cases}$$

and is given by

$$w_\varepsilon(t) = \gamma \int_0^t \exp\{\gamma \int_s^t u^*(\tau)d\tau\}v(s)[x^{u^*+\varepsilon v}(s) - x^{u^*}(s)]ds, \quad t \in [0,T].$$

By taking into account the first convergence we deduce that

$$w_\varepsilon \longrightarrow 0 \quad \text{in } C([0,T]),$$

and consequently

$$\frac{x^{u^*+\varepsilon v} - x^*}{\varepsilon} \longrightarrow z \quad \text{in } C([0,T]).$$

By (2.13) we obtain now that

$$\int_0^T [(1-u^*(t))z(t) - v(t)x^*(t)]dt \le 0. \tag{2.14}$$

Let us denote by p the solution to

$$\begin{cases} p'(t) = -\gamma u^*(t)p(t) + u^*(t) - 1, & t \in (0,T) \\ p(T) = 0. \end{cases} \tag{2.15}$$

p is given by

$$p(t) = -\int_t^T \exp\{\int_t^s \gamma u^*(\tau)d\tau\}(u^*(s) - 1)ds, \quad t \in [0,T].$$

If we multiply the differential equation in (2.15) by z and integrate over $[0,T]$ we get that

$$\int_0^T p'(t)z(t)dt = -\int_0^T \gamma u^*(t)p(t)z(t)dt + \int_0^T (u^*(t) - 1)z(t)dt.$$

If we integrate by parts it follows by ((2.11) and (2.15)) that

$$-\int_0^T p(t)z'(t)dt = -\int_0^T \gamma u^*(t)p(t)z(t)dt + \int_0^T (u^*(t) - 1)z(t)dt.$$

We again use (2.11) to obtain

$$-\int_0^T \gamma u^*(t)z(t)p(t)dt - \int_0^T \gamma v(t)x^*(t)p(t)dt$$
$$= -\int_0^T \gamma u^*(t)p(t)z(t)dt + \int_0^T (u^*(t) - 1)z(t)dt,$$

which implies

$$\int_0^T (1 - u^*(t))z(t)dt = \int_0^T \gamma v(t)x^*(t)p(t)dt.$$

This last relation and (2.14) imply that

$$\int_0^T x^*(t)(\gamma p(t) - 1)v(t)dt \leq 0, \tag{2.16}$$

for any $v \in V$. This is equivalent to

$$(\gamma p - 1)x^* \in N_K(u^*).$$

If we take into account the structure of $N_K(u^*)$ we may conclude that

$$u^*(t) = \begin{cases} 0 \text{ if } \gamma p(t) - 1 < 0 \\ \\ 1 \text{ if } \gamma p(t) - 1 > 0, \end{cases} \tag{2.17}$$

a.e. $t \in (0, T)$.

Let us give a direct proof of (2.17) starting from (2.16).

Denote by

$$A = \{t \in (0, T); \; \gamma p(t) - 1 < 0\}.$$

We prove that $u^*(t) = 0$ a.e. on A.

Assume by contradiction that there exists $\tilde{A} \subset A$, with $meas(\tilde{A}) > 0$ (meas denotes the Lebesgue measure; see Appendix A.1.1) such that $u^*(t) > 0$ a.e. in $\tilde{A}$. We can choose $v \in L^2(0, T)$ such that $v(t) < 0$ a.e. in $\tilde{A}$, $v(t) = 0$ a.e. in $(0, T) \setminus \tilde{A}$ and $0 \leq u^*(t) + \varepsilon v(t) \leq 1$ a.e. in $(0, T)$. It follows that

$$\int_0^T x^*(t)(\gamma p(t) - 1)v(t)dt = \int_{\tilde{A}} x^*(t)(\gamma p(t) - 1)v(t)dt > 0,$$

because $v(t) < 0$, $\gamma p(t) - 1 < 0$, $x^*(t) > 0$ on $\tilde{A}$, and $meas(\tilde{A}) > 0$. This is, of course, in contradiction to (2.16).

In the same manner it follows that

$$u^*(t) = 1 \text{ a.e. } t \in \{s \in (0, T); \; \gamma p(s) - 1 > 0\}.$$

The conclusion follows.

Remark 2.1. Equations (2.6), (2.15), and (2.17) represent the maximum principle and (2.15) and (2.17) are the first-order necessary optimality conditions for (P2).

Calculation of the optimal control u^*

Our next goal is to use Pontryagin's principle in order to get more information on the optimal control u^*. We show that for our particular problem we are able to calculate it exactly.

Let $(T - \eta, T]$ $(\eta > 0)$ be a maximal interval where the continuous function p satisfies $\gamma p(t) < 1$. By (2.17) and (2.15) we see that

$$p'(t) = -1, \quad t \in [T - \eta, T],$$

which implies that

$$p(t) = T - t \quad t \in [T - \eta, T].$$

Therefore, if $\gamma T > 1$ we have

$$p(t) = T - t \quad t \in [T - \frac{1}{\gamma}, T]$$

and

$$u^*(t) = 0 \ \text{ a.e. } t \in (T - \frac{1}{\gamma}, T).$$

Because $p(T - (\frac{1}{\gamma})) = \frac{1}{\gamma}$, we see that $p'(t) \leq 0$ on a maximal interval $(T - (1/\gamma) - \delta, T - (1/\gamma)]$ $(\delta > 0)$, and therefore $\gamma p(t) > 1$ on this interval. It also follows that

$$\begin{cases} p'(t) = -\gamma p(t) \\ u^*(t) = 1 \end{cases} \quad \text{on } (T - \frac{1}{\gamma} - \delta, T - \frac{1}{\gamma}).$$

Consequently

$$p(t) = \frac{1}{\gamma} \exp\{\gamma(T - \frac{1}{\gamma} - t)\} \quad t \in [T - \frac{1}{\gamma} - \delta, T - \frac{1}{\gamma}].$$

This implies that $\delta = T - (1/\gamma)$ and that $u^*(t) = 1$ a.e. $t \in [0, T - (1/\gamma))$.

The conclusion is that

- If $\gamma T > 1$, then

$$u^*(t) = \begin{cases} 1 & \text{if } t \in [0, T - \frac{1}{\gamma}) \\ 0 & \text{if } t \in [T - \frac{1}{\gamma}, T]; \end{cases} \quad (2.18)$$

- If $\gamma T \leq 1$, then

$$u^*(t) = 0, \quad t \in [0, T]. \quad (2.19)$$

This means that if the time interval is sufficiently long, then for a certain interval of time the rate of investment should be maximal. After that we do not invest any more (we just put everything for consumption).

A control u^* that takes values in a finite set $\{\alpha_1, \alpha_2, \ldots, \alpha_k\}$, and $(u^*)^{-1}(\alpha_i)$ is a measurable set for any $i \in \{1, 2, \ldots, k\}$ is called a bang-bang control.

If there exist $t_0 < t_1 < \cdots < t_k$ such that u^* is constant on any interval (t_{i-1}, t_i) $(i = \overline{1, k})$, then u^* is a bang-bang control on (t_0, t_k) and $t_1, t_2, \ldots, t_{k-1}$ are called switching points.

Remark 2.2. (i) The optimal control in our example is a bang-bang control and has at most one switching point, namely $T - (1/\gamma)$.

(ii) For our example we were able to calculate the optimal control. The form of the optimal control is given by (2.18) and (2.19). This is, of course, a fortunate situation.

(iii) After identifying $\mathcal{L}$, G, φ, f, and K we were able to write Pontryagin's principle formally. What we have done in this section was to prove it and use it in order to calculate the optimal control.

2.3 Maximizing the total population in a predator–prey system

The following Lotka–Volterra system,

$$\begin{cases} x'(t) = r_1 x(t) - \mu_1 x(t) y(t), & t \in (0, T) \\ y'(t) = -r_2 y(t) + \mu_2 x(t) y(t), & t \in (0, T) \end{cases}$$

$(T > 0)$ describes the dynamics of a predator–prey system on the time interval $(0, T)$. Here $x(t)$ represents the density of the prey population at moment t, and $y(t)$ the density of predators at moment t.

- $r_1 > 0$ is the intrinsic growth rate of prey in the absence of predators.
- $r_2 > 0$ is the decay rate of the predator population in the absence of prey.
- μ_1 and μ_2 are positive constants; $\mu_1 y(t)$ is the additional mortality rate of prey at moment t, due to predation (it is proportional to the predator population density); and $\mu_2 x(t)$ is the additional growth rate of prey at moment t, due to the presence of prey (it is proportional to the prey population density).

A more general model for the predator–prey system has been presented in Section 1.7.

If the prey are partially separated from predators then the functional response to predation changes and the system becomes

$$\begin{cases} x'(t) = r_1 x(t) - \mu_1 u(t) x(t) y(t), & t \in (0, T) \\ y'(t) = -r_2 y(t) + \mu_2 u(t) x(t) y(t), & t \in (0, T), \end{cases} \tag{2.20}$$

where $1 - u(t)$ represents the segregation rate at moment t $(0 \leq u(t) \leq 1)$.

Let the initial conditions be

$$\begin{cases} x(0) = x_0 > 0 \\ y(0) = y_0 > 0. \end{cases} \qquad (2.21)$$

We are interested in maximizing the total number of individuals of both populations at moment $T > 0$. The problem may be reformulated (see [Y82] and [Bar94]):

$$\text{Maximize}\{x^u(T) + y^u(T)\}, \qquad \textbf{(P3)}$$

subject to $u \in L^2(0,T)$, $0 \le u(t) \le 1$ a.e. $t \in (0,T)$, where (x^u, y^u) is the solution to (2.20) and (2.21).

Problem (P3) is a particular case of (P1), for $m = 1$, $N = 2$,

$$G(t, u, (x, y)) = 0,$$

$$\varphi(x, y) = x + y,$$

$$f(t, u, (x, y)) = \begin{pmatrix} r_1 x - \mu_1 u x y \\ -r_2 y + \mu_2 u x y \end{pmatrix},$$

and

$$K = \{w \in L^2(0,T); \ 0 \le w(t) \le 1 \text{ a.e. } t \in (0,T)\}.$$

Existence of an optimal pair for (P3)

Define

$$\Phi(u) = x^u(T) + y^u(T), \quad u \in K,$$

and let

$$d = \sup_{u \in K} \Phi(u).$$

It is obvious that $d \in [0, +\infty)$. For any $n \in \mathbb{N}^*$, there exists $u_n \in K$ such that

$$d - \frac{1}{n} < \Phi(u_n) \le d.$$

Because

$$x^{u_n}(t) = x_0 \exp\left\{ \int_0^t (r_1 - \mu_1 u(s) y^{u_n}(s)) ds \right\} > 0,$$

$$y^{u_n}(t) = y_0 \exp\left\{ \int_0^t (-r_2 + \mu_2 u(s) x^{u_n}(s)) ds \right\} > 0,$$

for $t \in [0,T]$, we get that $x^{u_n}(t), y^{u_n}(t) > 0$ for any $t \in [0,T]$, and so

$$0 \le (x^{u_n})'(t) \le r_1 x^{u_n}(t) \quad \text{a.e. } t \in (0,T).$$

This implies that
$$0 \le x^{u_n}(t) \le x_0 e^{r_1 T}, \quad t \in [0, T],$$
and that $\{(x^{u_n})'\}_n$ is bounded in $L^\infty(0, T)$.

On the other hand we get that
$$0 \le y^{u_n}(t) \le y_0 \exp\{(-r_2 + \mu_2 x_0 e^{r_1 T})T\}, \quad t \in [0, T],$$

and as a consequence $\{(y^{u_n})'\}_n$ is bounded in $L^\infty(0, T)$. It follows that $\{x^{u_n}\}_n$ and $\{y^{u_n}\}_n$ are bounded in $C([0, T])$, and uniformly equicontinuous. By Arzelà's theorem, and by taking into account that $\{u_n\}_n$ is bounded in $L^2(0, T)$ we get that on a subsequence we have

$$
\begin{aligned}
u_{n_k} &\longrightarrow u^* \text{ weakly in } L^2(0, T) \\
x^{u_{n_k}} &\longrightarrow x^* \text{ in } C([0, T]) \\
y^{u_{n_k}} &\longrightarrow y^* \text{ in } C([0, T])
\end{aligned}
\tag{2.22}
$$

($u^* \in K$ because K is a closed convex subset of $L^2(0, T)$, and consequently weakly closed).

Inasmuch as

$$x^{u_{n_k}}(t) = x_0 + \int_0^t [r_1 x^{u_{n_k}}(s) - \mu_1 u_{n_k}(s) x^{u_{n_k}}(s) y^{u_{n_k}}(s)] ds,$$

$$y^{u_{n_k}}(t) = y_0 + \int_0^t [-r_2 y^{u_{n_k}}(s) + \mu_2 u_{n_k}(s) x^{u_{n_k}}(s) y^{u_{n_k}}(s)] ds,$$

for any $t \in [0, T]$, and by taking into account (2.22) we get that

$$x^*(t) = x_0 + \int_0^t [r_1 x^*(s) - \mu_1 u^*(s) x^*(s) y^*(s)] ds,$$

$$y^*(t) = y_0 + \int_0^t [-r_2 y^*(s) + \mu_2 u^*(s) x^*(s) y^*(s)] ds,$$

for any $t \in [0, T]$, which means that (x^*, y^*) is the solution to (2.20) and (2.21) corresponding to u^* (i.e., $x^* = x^{u^*}$ and $y^* = y^{u^*}$). On the other hand by

$$d - \frac{1}{n_k} < x^{u_{n_k}}(T) + y^{u_{n_k}}(T) \le d \quad \text{for any } k \in \mathbb{N}^*,$$

and by using the convergences in (2.22) we may pass to the limit and obtain that

$$d = x^{u^*}(T) + y^{u^*}(T);$$

that is, u^* is an optimal control for (P3); $((u^*, (x^*, y^*))$ is an optimal pair for (P3); i.e., u^* is an optimal control and $x^* = x^{u^*}$, $y^* = y^{u^*}$).

The maximum principle for (P3)

For an arbitrary but fixed $v \in V = \{w \in L^2(0,T); \ u^* + \varepsilon w \in K$ for any $\varepsilon > 0$ sufficiently small$\}$ we consider (z_1, z_2) the solution to

$$\begin{cases} z_1' = r_1 z_1 - \mu_1 u^* z_1 y^* - \mu_1 u^* x^* z_2 - \mu_1 v x^* y^*, & t \in (0,T) \\ z_2' = -r_2 z_2 + \mu_2 u^* z_1 y^* + \mu_2 u^* x^* z_2 + \mu_2 v x^* y^*, & t \in (0,T) \\ z_1(0) = z_2(0) = 0. \end{cases} \quad (2.23)$$

Because

$$x^*(T) + y^*(T) \geq x^{u^* + \varepsilon v}(T) + y^{u^* + \varepsilon v}(T),$$

we get that

$$\frac{x^{u^* + \varepsilon v}(T) - x^*(T)}{\varepsilon} + \frac{y^{u^* + \varepsilon v}(T) - y^*(T)}{\varepsilon} \leq 0, \quad (2.24)$$

for any $\varepsilon > 0$ sufficiently small.

For $\varepsilon > 0$ sufficiently small we have that $x^{u^* + \varepsilon v}$ satisfies

$$(x^{u^* + \varepsilon v})'(t) \leq r_1 x^{u^* + \varepsilon v}(t) \text{ a.e. } t \in (0,T),$$

and consequently it follows that there exists $M \in (0, +\infty)$ such that

$$0 \leq x^{u^* + \varepsilon v}(t) \leq M \text{ for any } t \in [0,T],$$

for any $\varepsilon > 0$ sufficiently small. On the other hand

$$(y^{u^* + \varepsilon v})'(t) \leq (-r_2 + M\mu_2) y^{u^* + \varepsilon v}(t) \text{ a.e. } t \in (0,T),$$

and this implies that $\{y^{u^* + \varepsilon v}\}$ is bounded in $C([0,T])$ (for $\varepsilon > 0$ sufficiently small). It follows that both sequences $\{x^{u^* + \varepsilon v}\}$ and $\{y^{u^* + \varepsilon v}\}$ are uniformly bounded and uniformly equicontinuous on $[0,T]$. By Arzelà's theorem it follows that on a sequence $\varepsilon_n \searrow 0$ we have that

$$x^{u^* + \varepsilon_n v} \longrightarrow \tilde{x} \text{ in } C([0,T]),$$
$$y^{u^* + \varepsilon_n v} \longrightarrow \tilde{y} \text{ in } C([0,T]). \quad (2.25)$$

Because

$$x^{u^* + \varepsilon_n v} = x_0 + \int_0^t [r_1 x^{u^* + \varepsilon_n v}(s) - \mu_1(u^*(s) + \varepsilon_n v(s)) x^{u^* + \varepsilon_n v}(s) y^{u^* + \varepsilon_n v}(s)] ds$$

and

$$y^{u^* + \varepsilon_n v} = y_0 + \int_0^t [-r_2 y^{u^* + \varepsilon_n v}(s) + \mu_2(u^*(s) + \varepsilon_n v(s)) x^{u^* + \varepsilon_n v}(s) y^{u^* + \varepsilon_n v}(s)] ds,$$

for any $t \in [0, T]$, we pass to the limit (and use (2.25)), and we get

$$\tilde{x}(t) = x_0 + \int_0^t [r_1\tilde{x}(s) - \mu_1 u^*(s)\tilde{x}(s)\tilde{y}(s)]ds,$$

and

$$\tilde{y}(t) = y_0 + \int_0^t [-r_2\tilde{y}(s) + \mu_2 u^*(s)\tilde{x}(s)\tilde{y}(s)]ds,$$

for any $t \in [0, T]$, which means that $(\tilde{x}, \tilde{y})$ is the solution to (2.20) corresponding to u^*; that is, $\tilde{x} = x^{u^*}$, $\tilde{y} = y^{u^*}$.

Define now

$$\alpha_n(t) = \frac{1}{\varepsilon_n}\left[x^{u^* + \varepsilon_n v}(t) - x^*(t)\right] - z_1(t), \quad t \in [0, T],$$

$$\beta_n(t) = \frac{1}{\varepsilon_n}\left[y^{u^* + \varepsilon_n v}(t) - y^*(t)\right] - z_2(t), \quad t \in [0, T].$$

(α_n, β_n) is the solution to

$$\begin{cases} \alpha_n' = r_1\alpha_n - \mu_1 u^*\alpha_n y^* - \mu_1 u^* x^*\beta_n + f_{1n}(t), & t \in (0, T) \\ \beta_n' = -r_2\beta_n + \mu_2 u^*\alpha_n y^* + \mu_2 u^* x^*\beta_n + f_{2n}(t), & t \in (0, T) \\ \alpha_n(0) = \beta_n(0) = 0 \end{cases}$$

and $f_{1n} \longrightarrow 0$, $f_{2n} \longrightarrow 0$ in $L^\infty(0, T)$.

This yields

$$\alpha_n(t)^2 + \beta_n(t)^2 \le c \int_0^t [\alpha_n(s)^2 + \beta_n(s)^2]ds$$

$$+ 2\int_0^t [f_{1n}(s)\alpha_n(s) + f_{2n}(s)\beta_n(s)]ds$$

$$\le (c+1)\int_0^t [\alpha_n(s)^2 + \beta_n(s)^2]ds$$

$$+ \int_0^T [f_{1n}(t)^2 + f_{1n}(t)^2]dt,$$

$t \in [0, T]$, where $c > 0$ is a constant independent of n. By Bellman's lemma (see Appendix A.2) we conclude that

$$0 \le \alpha_n(t)^2 + \beta_n(t)^2 \le e^{(c+1)t}\int_0^T [f_{1n}(t)^2 + f_{1n}(t)^2]dt,$$

for any $t \in [0, T]$. We pass to the limit and conclude that

$$\alpha_n \longrightarrow 0, \quad \beta_n \longrightarrow 0 \text{ in } C([0, T]).$$

This implies that

$$\frac{1}{\varepsilon_n}[x^{u^* + \varepsilon_n v} - x^*] \longrightarrow z_1 \text{ in } C([0, T]),$$

and

$$\frac{1}{\varepsilon_n}[y^{u^* + \varepsilon_n v} - y^*] \longrightarrow z_2 \text{ in } C([0, T]).$$

If we again use (2.24) we may infer that

$$z_1(T) + z_2(T) \le 0. \tag{2.26}$$

Let (p_1, p_2) be the solution to

$$\begin{cases} p_1' = -r_1 p_1 + \mu_1 u^* y^* p_1 - \mu_2 u^* y^* p_2, \ t \in (0, T) \\ p_2' = r_2 p_2 + \mu_1 u^* x^* p_1 - \mu_2 u^* x^* p_2, \quad t \in (0, T) \\ p_1(T) = p_2(T) = 1. \end{cases} \tag{2.27}$$

By multiplying the first equation in (2.27) by z_1 and the second one by z_2 and integrating over $[0, T]$ we get that

$$\int_0^T [p_1'(t) z_1(t) + p_2'(t) z_2(t)] dt$$
$$= \int_0^T [-r_1 p_1(t) z_1(t) + \mu_1 u^*(t) y^*(t) p_1(t) z_1(t) - \mu_2 u^*(t) y^*(t) p_2(t) z_1(t)$$
$$+ \mu_1 u^*(t) x^*(t) p_1(t) z_2(t) - \mu_2 u^*(t) x^*(t) p_2(t) z_2(t) + r_2 p_2(t) z_2(t)] dt.$$

If we integrate by parts and use (2.23) we get after some calculation that

$$p_1(T) z_1(T) + p_2(T) z_2(T) - p_1(0) z_1(0) - p_2(0) z_2(0)$$
$$= \int_0^T x^*(t) y^*(t) v(t) [\mu_2 p_2(t) - \mu_1 p_1(t)] dt,$$

and consequently by (2.23) and (2.26) we get that

$$z_1(T) + z_2(T) = \int_0^T x^*(t) y^*(t) v(t) [\mu_2 p_2(t) - \mu_1 p_1(t)] dt \le 0,$$

for any $v \in V$. This implies (as in the previous section) that

$$u^*(t) = \begin{cases} 0 \text{ if } x^*(t) y^*(t) [\mu_2 p_2(t) - \mu_1 p_1(t)] < 0 \\ 1 \text{ if } x^*(t) y^*(t) [\mu_2 p_2(t) - \mu_1 p_1(t)] > 0 \end{cases}$$

a.e. on $(0, T)$. Because x_0, $y_0 > 0$, and x^* and y^* are positive functions, we may conclude that

$$u^*(t) = \begin{cases} 0 \text{ if } \mu_2 p_2(t) - \mu_1 p_1(t) < 0 \\ 1 \text{ if } \mu_2 p_2(t) - \mu_1 p_1(t) > 0 \end{cases} \tag{2.28}$$

a.e. on $(0, T)$.

Equations (2.27) and (2.28) are the first-order necessary optimality conditions, and (2.20)–(2.21), (2.27)–(2.28) represent the maximum principle for (P3).

The structure of the optimal control u^* for (P3)

Our next goal is to obtain more information about the structure of the optimal control u^*.

- If $\mu_2 < \mu_1$, then $\mu_2 p_2(T) - \mu_1 p_1(T) = \mu_2 - \mu_1 < 0$, and then we may choose a maximal interval $(T - \eta, T]$ $(\eta > 0)$ where $\mu_2 p_2(t) - \mu_1 p_1(t) < 0$. By (2.28) we have $u^*(t) = 0$ on $(T - \eta, T]$ and consequently

$$p_1'(t) = -r_1 p_1(t), \quad p_2'(t) = r_2 p_2(t) \text{ a.e. } t \in (T - \eta, T).$$

This yields

$$p_1(t) = \exp\{-r_1(t - T)\}, \quad p_2(t) = \exp\{r_2(t - T)\}, \quad t \in [T - \eta, T].$$

The function $t \mapsto \mu_2 \exp\{r_2(t - T)\} - \mu_1 \exp\{-r_1(t - T)\}$ is increasing on $[T - \eta, T]$, and this implies that $T - \eta = 0$ and

$$\mu_2 p_2(t) - \mu_1 p_1(t) < 0, \quad t \in (0, T),$$

and so $u^*(t) = 0$ a.e. on $(0, T)$.

- If $\mu_2 = \mu_1$, then (p_1, p_2) is the solution to

$$\begin{cases} p_1' = -r_1 p_1 - \mu_1 u^* y^* (p_2 - p_1), \, t \in (0, T) \\ p_2' = r_2 p_2 - \mu_1 u^* x^* (p_2 - p_1), \quad t \in (0, T) \\ p_1(T) = p_2(T) = 1. \end{cases}$$

In conclusion

$$p_2(t) - p_1(t) = -\int_t^T [r_2 p_2(s) + r_1 p_1(s)] \exp\{\mu_1 \int_s^t u^*(\tau)[y^*(\tau) - x^*(\tau)]d\tau\} ds,$$

$t \in [0, T]$. So, $p_2(t) - p_1(t) < 0$ on a maximal interval $(T - \eta, T]$ $(\eta > 0)$ and, in the same manner as in the previous case, it follows that $u^*(t) = 0$ a.e. on $(0, T)$.

- If $\mu_2 > \mu_1$, then there exists a maximal interval $(T - \eta, T]$ $(\eta > 0)$ such that

$$\mu_2 p_2(t) - \mu_1 p_1(t) > 0, \quad t \in (T - \eta, T].$$

By (2.28) we have $u^*(t) = 1$ on $(T - \eta, T]$. We intend to prove that $T - \eta$ is a switching point for the optimal control u^*. Indeed, by (2.27) we get that

$$\mu_2 p_2(t) - \mu_1 p_1(t) = -\int_t^{T-\eta} [r_2 \mu_2 p_2(s) + r_1 \mu_1 p_1(s)] \tag{2.29}$$
$$\cdot \exp\{\int_s^t u^*(\tau)[\mu_2 x^*(\tau) - \mu_1 y^*(\tau)]d\tau\} ds,$$

$t \in [0, T - \eta]$. On the other hand (p_1, p_2) is a solution to

$$\begin{cases} p_1' = -p_1(r_1 - \mu_1 y^*) - \mu_2 y^* p_2, \ t \in (T - \eta, T) \\ p_2' = -p_2(\mu_2 x^* - r_2) + \mu_1 x^* p_1, \ t \in (T - \eta, T) \\ p_1(T) = p_2(T) = 1. \end{cases} \qquad (2.30)$$

Because $\mu_2 p_2(t) - \mu_1 p_1(t) > 0$, for any $t \in (T - \eta, T]$, then we get that

$$p_1(t) \geq \exp\{r_1(T - t)\} \geq 1, \quad t \in [T - \eta, T].$$

Using the fact that $\mu_2 p_2(T - \eta) - \mu_1 p_1(T - \eta) = 0$ and (2.29) we obtain that $p_2(T - \eta) > 0$ and consequently $\mu_2 p_2(t) - \mu_1 p_1(t) < 0$ in a maximal interval $(T - \eta - \varepsilon, T - \eta]$ ($\varepsilon > 0$). This implies that $u^*(t) = 0$ on $(T - \eta - \varepsilon, T - \eta]$. On this interval we have

$$p_1(t) = p_1(T - \eta)\exp\{r_1(T - \eta - t)\},$$
$$p_2(t) = p_2(T - \eta)\exp\{r_2(t - T + \eta)\},$$

and in conclusion $\mu_2 p_2 - \mu_1 p_1$ is increasing on $(T - \eta - \varepsilon, T - \eta)$. Hence

$$\mu_2 p_2(t) - \mu_1 p_1(t) < 0, \quad t \in (T - \eta - \varepsilon, T - \eta),$$

and consequently $T - \eta - \varepsilon = 0$. The conclusion is that

$$u^*(t) = \begin{cases} 0, & t \in [0, T - \eta] \\ 1, & t \in (T - \eta, T] \end{cases} \qquad (2.31)$$

a.e. on $(0, T)$.

So, we have a bang-bang optimal control with at most one switching point. We can determine the switching point $T - \eta$, either by taking into account (2.30) and $\mu_2 p_2(T - \eta) - \mu_1 p_1(T - \eta) = 0$, or by finding $T - \eta \in [0, T]$, which maximizes $\Phi(u^*)$, where u^* is given by (2.31).

Approximating the optimal control for (P3)

In order to approximate the optimal control u^* we have to find η from formula (2.31). A simple idea is to try τ ($T - \eta$ in (2.31)) as switching point for the control of the elements of a grid defined on $[0, L]$ (we put L instead of T) and to get the one that provides the maximum value for $\Phi(u)$. Here

$$u(t) = \begin{cases} 0, & t \in [0, \tau] \\ 1, & t \in (\tau, L]. \end{cases}$$

Here is the algorithm.

Algorithm 2.1

```
/* Build the grid */
tspan = 0:h1:L ;
/* Try the grid points */
m = length(tspan) ;
```

for i = 1 to m
    τ = tspan(i) ;
    /* **S1** : Build the corresponding control u_τ */

$$u_\tau(t) = \begin{cases} 0, & t \in [0,\tau] \\ 1, & t \in (\tau, L]. \end{cases}$$

/* **S2** : Compute the state $[x,y]$, the corresponding solution of system (2.20) corresponding to $u := u_\tau$, with the initial conditions */

$$x(0) = x_0, \quad y(0) = y_0.$$

/* **S3** : Compute the corresponding value of the cost functional Φ */
fiu(i) = x(L) + y(L) ;
end–for
/* **S4** : Find the maximal value of vector fiu */

Here is the corresponding program.

```
% file ppp1.m
% predator–prey model with bang-bang optimal control
clear
global r1 r2 mu1 mu2
global tsw
disp('get model parameters') ;
r1 = input('r1 : ') ;
mu1 = input('mu1 : ') ;
r2 = input('r2 : ') ;
mu2 = input('mu2 : ') ;
disp('get data') ;
L = input('final time : ') ;
h = input('grid step : ') ;
h1 = input('switch step : ') ;
x0 = input('x(0) : ') ;
y0 = input('y(0) : ') ;
lw = input('LineWidth : ') ; % for graphs ( plot )
tt = 0:h:L ; % ODE integration grid
n = length(tt) ;
tspan = 0:h1:L ; % switching points grid
m = length(tspan) ;
for i = 1:m
    i
    tsw = tspan(i) ; % tsw stands for switching point τ
    [t q] = ode45('bp2',tt,[x0 ; y0]) ;
    k = length(t) ;
    fiu(i) = q(k,1) + q(k,2) ; % store cost functional value
    clear t q % clear memory to avoid garbage for the next iteration
```

```
end
w = fiu' ;
save cont.txt w -ascii
disp('FILE MADE') ;
[vmax,j] = max(fiu) ; % maximal value and corresponding index
j
a1 = ['max = ', num2str(vmax)] ;
disp(a1) ;
a2 = ['switch = ', num2str(tspan(j))] ;
disp(a2) ;
plot(tspan,fiu,'LineWidth',lw) ; grid
xlabel('\bf u switch','FontSize',16)
ylabel('\bf \Phi(u_{\tau})','FontSize',16)
figure(2)
bar(fiu)
title('\bf \Phi(u_{\tau})','FontSize',16)
```

We have used a vector, namely *tt*, for *ode45* and another one, namely *tspan*, for the switching points grid to get a faster program.

Here is the function file *bp2.m* for the right-hand side of the differential system.

```
function out1 = bp2(t,q)
global r1 r2 mu1 mu2
global tsw
if t > tsw
    u = 1 ;
else
    u = 0 ;
end
out1 = [ r1*q(1) - mu1*u*q(1)*q(2) ; mu2*u*q(1)*q(2) - r2*q(2) ] ;
```

For a numerical test we have used $r_1 = 0.07$, $\mu_1 = 1$, $r_2 = 0.6$, $\mu_2 = 2$, $L = 50$, $h = 0.1$, $h1 = 1$, $x(0) = 0.04$, $y(0) = 0.02$, and $lw = 5$. The graph of the corresponding function $\tau \mapsto \Phi(u_\tau)$, where τ is the switching point of u_τ, can be seen in Figures 2.1 and 2.2.
We have obtained a global maximum on $[0, L]$ for $\tau^* = 15$, and the maximal value of the cost functional is 1.5006. The program that uses the switch point of the optimal control in order to plot the graphs for the corresponding state components is *ppp2.m*:

```
% file ppp2.m
% predator–prey model with bang-bang optimal control
% makes graphs by using the switching point obtained by ppp1.m
clear
global r1 r2 mu1 mu2
global tsw
```

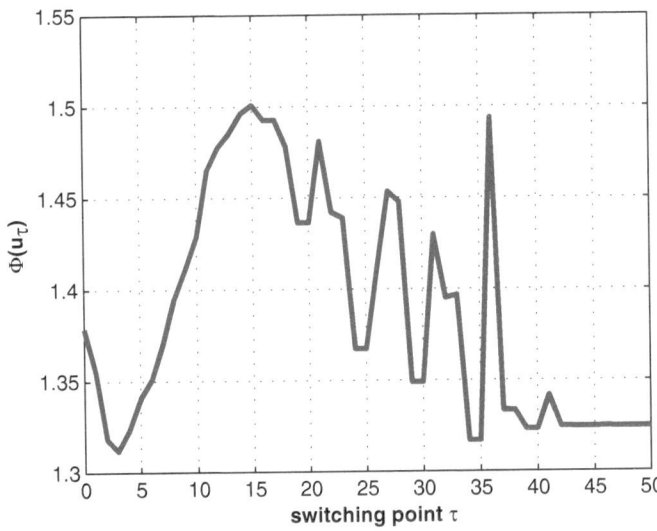

Fig. 2.1. The dependence of the cost function with respect to the switching point

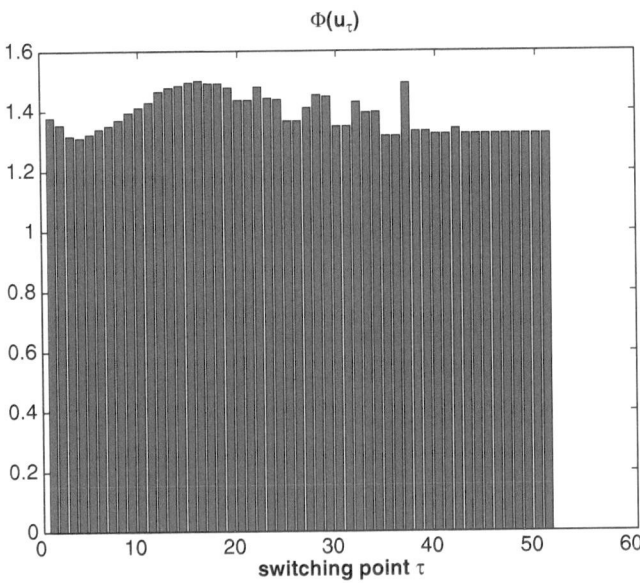

Fig. 2.2. Another representation of the dependence of the cost function with respect to the switching point τ

... read parameters and data as in ppp1.m (except h) ...
```
tspan = 0:h1:L ;
% graph of the control
m = length(tspan) ;
for i = 1:m
    if tspan(i) > tsw
        z(i) = 1 ;
    else
        z(i) = 0 ;
    end
end
plot(tspan,z,'rs') ; grid
axis([0 L −0.2 1.2])
xlabel('\bf t','Fontsize',16)
ylabel('\bf u(t)','Fontsize',16)
[t q] = ode45('bp2',[0 L],[x0 ; y0]) ;
% predator–prey populations graph
figure(2)
plot(t,q(:,1),'*',t,q(:,2),'ro') ; grid
xlabel('\bf t','FontSize',16)
legend('prey','predator',0)
% xOy graph
figure(3)
plot(q(:,1),q(:,2),'LineWidth',lw) ; grid
xlabel('\bf x','Fontsize',16)
ylabel('\bf y','Fontsize',16)
```

2.4 Insulin treatment model

We consider a model for insulin treatment for patients with diabetes. The main problem for such a patient is to keep the blood glucose level close to a convenient value and to avoid large variations of it. In practice insulin injections are used. An optimal control problem with impulsive controls is considered to maintain a steady state of the blood glucose level. This problem does not fit in the framework of Problem (P1) from Section 2.1 mainly because the control considered here is of impulsive type. We do, however, obtain first-order necessary optimality conditions which are used to write a program.

In the case of diabetes the pancreas (the beta cells) is not able to provide enough insulin to metabolize glucose. Blood glucose concentration increases when glucose is administrated in mammals whereas insulin accelerates the removal of glucose from the plasma. Therefore blood sugar decays to a normal value of 0.8–$1.2\,g/l$. Let us denote by $I(t)$ the insulin concentration, and by $G(t)$ the glucose concentration at moment $t \in [0, L]$ ($L > 0$).

We now consider diabetic patients who are not able to produce enough insulin. The insulin is supplied by injections. The glucose concentration can be easily determined (measured). A corresponding simplified model for dynamics of the insulin–glucose system is the following one (see [Che86, Chapter 6]):

$$
\begin{cases}
I'(t) = dI(t), & t \in (0, L) \\
G'(t) = bI(t) + aG(t), & t \in (0, L) \\
I(0) = I_0, \quad G(0) = G_0,
\end{cases}
\tag{2.32}
$$

where $d < 0$ ($|d|$ is the decay rate of insulin), a is the growth rate of glucose ($a \neq d$), b is a negative constant that can be measured, I_0 is the initial concentration of insulin (injected), and G_0 is the initial concentration of glucose. The numerical tests show that model (2.32) works well only for $I(t)$ and $G(t)$ between appropriate limits. For I_0 and G_0 outside the usual medical limits it is possible to obtain negative values for $I(t)$ and $G(t)$ and therefore the model fails. A more accurate model is, however, indicated at the end of this subsection. The reaction between $I(t)$ and $G(t)$ in (2.32) is a local linearization of the full model presented later (see (2.41)).

The first program plots the graphs of insulin concentration and of glucose concentration.

```
% file dbt1.m
% blood insulin–glucose system
% y(1) = insulin concentration
% y(2) = glucose concentration
clear
global a b d
L = input('final time : ') ;
h = input('h : ') ;
I0 = input('I(0) : ') ;
G0 = input('G(0) : ') ;
a = 0.0343 ;
b = -0.05 ;
d = -0.5 ;
tspan = 0:h:L ;
[t y] = ode45('hum1',tspan,[I0 ; G0]) ;
plot(t,y(:,1),'*') ; grid
xlabel('\bf t','FontSize',16)
ylabel('\bf I(t)','FontSize',16)
figure(2)
plot(t,y(:,2),'r*') ; grid
xlabel('\bf t','FontSize',16)
ylabel('\bf G(t)','FontSize',16)
```

We also have

```
function out1 = hum1(t,y)
global a b d
out1 = [d*y(1) ; b*y(1) + a*y(2)] ;
```

The numerical test is done for $L = 10$, $h = 0.01$, $I0 = 15$, and $G0 = 2$. The evolution of insulin and glucose concentration are presented in Figures 2.3 and 2.4, respectively. Notice that $I(t)$ decays to zero (the effect of the decay rate) and $G(t)$ reaches a convenient level. The insulin has a good effect because the glucose level at the beginning was $G_0 = 2$, and reaches approximatively, the value 0.8 at the moment $t = 6$. After $t = 7$ the insulin effect almost vanishes and the glucose level increases slowly.

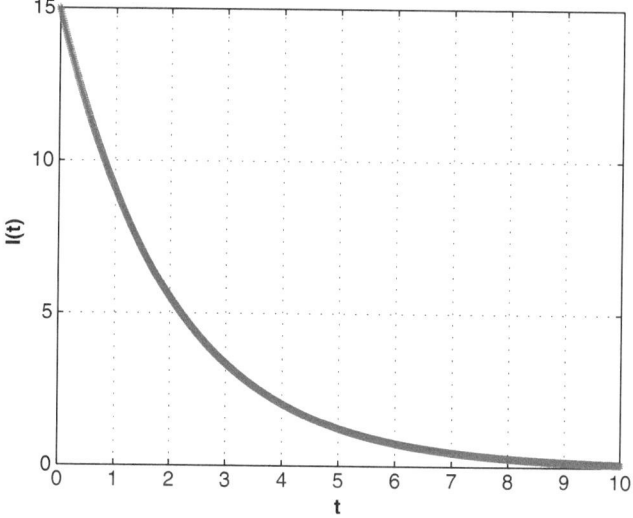

Fig. 2.3. Insulin dynamics

System (2.32) can also be integrated mathematically. We first consider the problem of insulin dynamics:

$$\begin{cases} I'(t) = dI(t), \ t \in (0, L) \\ I(0) = I_0 \end{cases}$$

which has a unique solution given by

$$I(t) = I_0 e^{dt}, \quad t \in [0, L]. \tag{2.33}$$

If we use the form of $I(t)$ given by (2.33), we obtain from (2.32) the following linear model for the glucose dynamics,

$$\begin{cases} G'(t) = bI_0 e^{dt} + aG(t), \quad t \in (0, L) \\ G(0) = G_0, \end{cases}$$

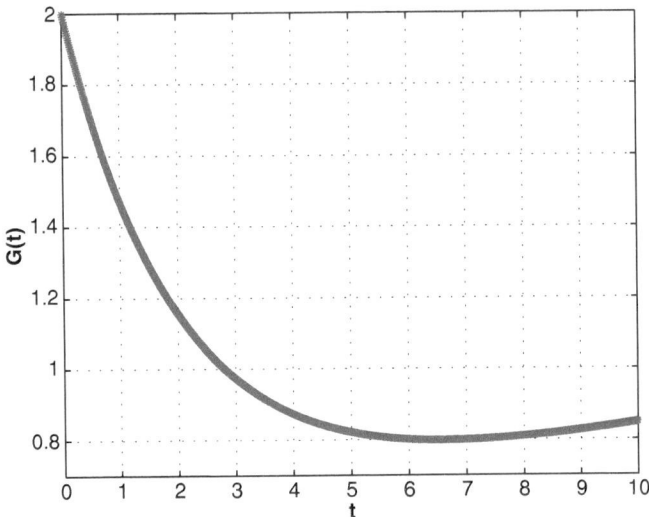

Fig. 2.4. Glucose dynamics

which gives the following formula for the glucose concentration,

$$G(t) = G_0 e^{at} + \frac{bI_0}{d-a}(e^{dt} - e^{at}), \quad t \in [0, L]. \tag{2.34}$$

Hence the solution of system (2.32) is given by formulae (2.33) and (2.34). The corresponding program is

```
% file dbt2.m
% blood insulin–glucose system
% y1(t) = insulin concentration
% y2(t) = glucose concentration
% mathematical integration
clear
L = input('final time : ') ;
h = input('h : ') ;
I0 = input('insulin(0) : ') ;
G0 = input('glucose(0) : ') ;
a = 0.0343 ;
b = -0.05 ;
d = -0.5 ;
temp = b*I0/(d - a) ;
t = 0:h:L ;
v = exp(d*t) ;
w = exp(a*t) ;
y1 = I0*v ;
```

y2 = G0*w + temp*(v − w) ;
% make figures as in previous program
% .....

The figures obtained are similar to the previous two figures.

We now consider an optimal control problem with impulsive control to obtain a scheme of insulin treatment providing good control of glycemia over some time interval. We denote by A the desired level of glucose. Assume that the patient gets m injections of insulin at moments

$$0 = t_1 < t_2 < \cdots < t_m = L,$$

with corresponding amounts $c_j = c(t_j)$, $j = 1, 2, \ldots, m$ and that the initial concentration of insulin is $I_0 = 0$. Usually the moments for injections are fixed and we have $t_{j+1} - t_j = h$ for $j = 1, \ldots, m - 1$. The dynamics of the insulin–glucose system is then described by

$$\begin{cases} I'(t) = dI(t) + \sum_{j=1}^{m} c_j \delta_{t_j}, \\[2em] G'(t) = bI(t) + aG(t), \\ I(0) = 0, \quad G(0) = G_0, \end{cases} \tag{2.35}$$

where δ_{t_j} is the Dirac mass at t_j. System (2.35) is equivalent to the following one

$$\begin{cases} I'(t) = dI(t), & t \in (t_j, t_{j+1}), j \in \{1, \ldots, m - 1\} \\ I(0) = 0 \\ I(t_j+) = I(t_j-) + c_j, & j \in \{1, \ldots, m - 1\} \\ G'(t) = bI(t) + aG(t), & t \in (0, L) \\ G(0) = G_0. \end{cases} \tag{2.36}$$

The solution of (2.35) in the sense of the theory of distributions (which is also the solution to (2.36)) is given by

$$\begin{cases} I(t) = \sum_{j=1}^{m} c_j H(t - t_j) e^{d(t-t_j)}, \\[2em] G(t) = G_0 e^{at} + \dfrac{b}{d-a} S(t), \end{cases} \tag{2.37}$$

$t \in [0, L]$, where

$$S(t) = \sum_{j=1}^{m} c_j H(t - t_j) \left[e^{d(t-t_j)} - e^{a(t-t_j)} \right], \tag{2.38}$$

and H is the step (Heaviside) function (i.e., $H : \mathbb{R} \to \mathbb{R}$),

$$H(t) = \begin{cases} 1 & \text{if } t \geq 0 \\ 0 & \text{if } t < 0. \end{cases}$$

Therefore, function $t \mapsto H(t - t_j)$ in formulae (2.37) and (2.38), defined for $t \in [0, L]$, reads

$$H(t - t_j) = \begin{cases} 1 & \text{if } t \in [t_j, L] \\ 0 & \text{if } t \in [0, t_j). \end{cases}$$

The formula for G says that the effect of the insulin injection received at the moment $t = t_j$ is valid only for $t \geq t_j$. The effect vanishes after some time due to the exponential function with negative exponent.

Here is the optimal control problem (the insulin treatment) related to (2.35):

$$\text{Minimize} \ \ \Psi(c) = \frac{1}{2} \int_0^L [G(t) - A]^2 dt, \tag{I}$$

subject to $c = (c_1, \ldots, c_m) \in \mathbb{R}^m$, where (I, G) is the solution to (2.35). Here the vector c is the control (which is in fact an impulsive control, a control that acts only at some discrete moments of time).

The functional Ψ is quadratic with respect to every c_j, thus it means that there exists at least an optimal control $c = (c_1, \ldots, c_m) \in \mathbb{R}^m$. The optimal control satisfies

$$\frac{\partial \Psi}{\partial c_j}(c) = 0, \quad j = 1, \ldots, m, \tag{2.39}$$

a linear algebraic system with the unknowns c_j, $j = 1, \ldots, m$. We calculate the partial derivatives and use formula (2.39) to get the following algebraic linear system,

$$\sum_{i=1}^m q_{ij} c_i = B_j, \quad j \in \{1, \ldots, m\},$$

where

$$q_{ij} = \alpha \int_0^L H(t - t_i) H(t - t_j) e_i(t) e_j(t) dt, \tag{2.40}$$

$$B_j = \int_0^L H(t - t_j) e_j(t)(A - G_0 e^{at}) dt = \int_{t_j}^L e_j(t)(A - G_0 e^{at}) dt,$$

$i, j \in \{1, \ldots, m\}$. We have denoted

$$\alpha = \frac{b}{d - a},$$

and

$$e_j(t) = e^{d(t - t_j)} - e^{a(t - t_j)}, \quad t \in [0, L], \ j \in \{1, \ldots, m\}.$$

If $i > j$, then $t_i > t_j$ and Formula (2.40) reads

$$q_{ij} = \alpha \int_{t_i}^{L} e_i(t)e_j(t)dt.$$

Our goal is to solve system (2.39). However if a certain component c_j is negative, this is meaningless from the medical point of view. If we introduce the restrictions $c_j \geq 0, j \in \{1, \ldots, m\}$, we get a mathematical programming problem which is more complicated. Another possibility is to introduce restrictions of the form $0 \leq c_j \leq \bar{c}, j \in \{1, \ldots, m\}$, and to use a projected gradient method (see Chapter 3). But this is more complicated also. To establish a treatment policy we can simply take $c_j := 0$ if $c_j < 0$. Then we have to add glucose, usually from food, or to replace the negative dose of the injection by $c_j = 0$, and consequently to obtain suboptimal control. For our numerical test made for medically appropriate values of $G(0)$ the solution was positive.

We return to the linear system. The algorithm to compute the transpose of matrix Q, that is, $Q^T = [q_{ij}]$, is:

for j = 1 to m
    for i = 1 to j
        compute $q_{ij} = \alpha \int_{t_j}^{L} e_i(t)e_j(t)dt$
    end–for
    for i = j+1 to m
        compute $q_{ij} = \alpha \int_{t_i}^{L} e_i(t)e_j(t)dt$
    end–for
end–for

Then we transpose the matrix $[q_{ij}]$ obtained above and we get Q. We leave it to the reader to write the corresponding program. The values of the system parameters are $a = 0.1$, $b = -0.05$, and $d = -0.5$. Below we give only the sequence to compute the matrix Q and the right-hand side B of the system $Qc = B$.

```
...
Q = zeros(m − 1) ;
for j = 1:m − 1
    tj = t(j) ;
    for i = 1:j
        ti = t(i) ;
        Q(i,j) = alf*quadl('fi1',tj,L) ;
    end
    for i = j+1:m − 1
        ti = t(i) ;
        Q(i,j) = alf*quadl('fi2',ti,L) ;
    end
end
```

```
Q = Q' ;
for j = 1:m − 1
    tj = t(j) ;
    B(j) = quadl('psi',t(j),L) ;
end
B = B' ;
% solve system Qc = B
c = Q\B ;
...
```

The function file $fi1.m$ computes the matrix components q_{ij} for $i \leq j$.

```
function y = fi1(t)
global ti tj
global a d
y = 0 ;
if t >= tj
    temp1 = exp(d*(t − tj)) − exp(a*(t − tj)) ;
    temp2 = exp(d*(t − ti)) − exp(a*(t − ti)) ;
    y = temp1 .* temp2 ;
end
```

The function file $fi2.m$ computes the matrix components q_{ij} for $i > j$. It is similar to $fi1.m$. There is only one difference. The statement

```
if t >= tj
```

is replaced by

```
if t >= ti
```

The function file $psi.m$ computes the right-hand side components B_j.

```
function y = psi(t)
global tj
global a d
global a1
global G0
y = 0 ;
if t >= tj
    temp1 = exp(d*(t − tj)) − exp(a*(t − tj)) ;
    temp2 = a1 − G0*exp(a*t) ;
    y = temp1 .* temp2 ;
end
```

We pass now to numerical examples.

Example 1. We take $L = 48$ (hours), $G(0) = 2$, $A = 1$, and $m = 9$ (number of injections). It follows that the interval between successive injections is $h = 6$ (hours). The insulin "shots" are represented in Figure 2.5.

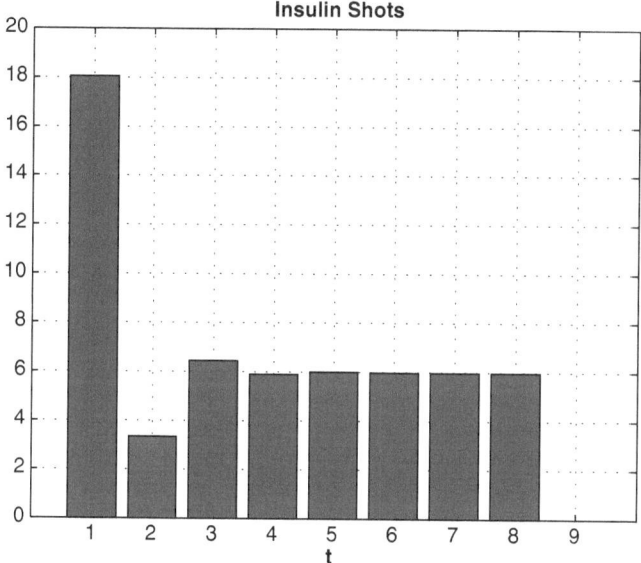

Fig. 2.5. The insulin doses for 48 h

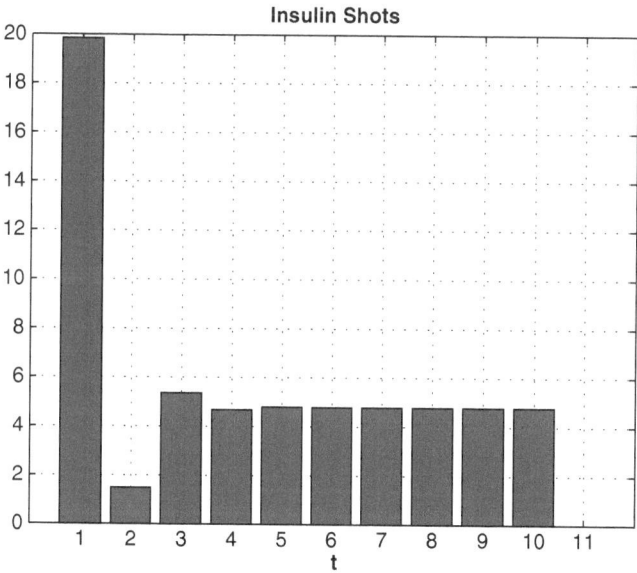

Fig. 2.6. The insulin doses for 60 h

Example 2. Another experiment was done with $L = 60$, $G(0) = 2$, $A = 0.8$, and $m = 11$ ($h = 6$). The results are given in Figure 2.6.

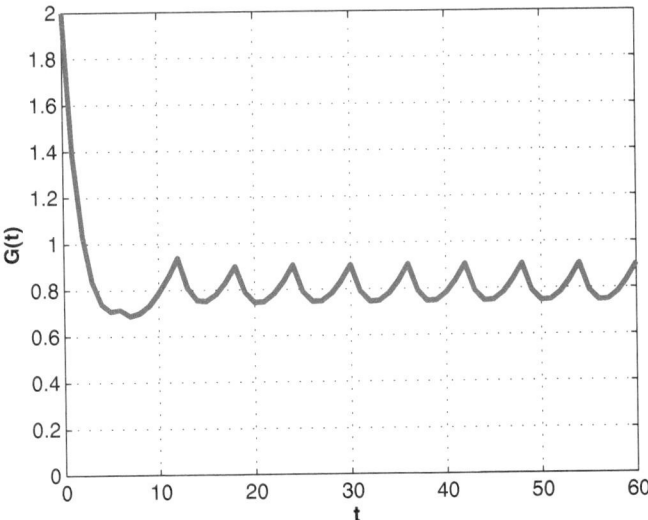

Fig. 2.7. The blood glucose concentration for the second numerical experiment

To complete our investigation we have also computed the glucose level given by formulae (2.37) and (2.38). The shape of the blood glucose concentration for the second numerical experiment is given in Figure 2.7. Let us remark that the glucose level decays from $G(0) = 2$ under the desired level $A = 0.8$ and then remains quite close to it. For the first numerical experiment the behavior of $G(t)$ is similar.

Remark 2.3. Equation (2.39) are the first-order optimality conditions for Problem (I).

We propose that the reader investigate in a similar manner the following optimal control problem.

$$\text{Minimize } \Psi(c) = \frac{1}{2} \int_0^L [G(t) - A]^2 dt, \tag{I1}$$

subject to $c = (c_1, c_2, \ldots, c_m) \in \mathbb{R}^m$, where (I, G) is the solution to the following more accurate model,

$$\begin{cases} I'(t) = dI(t) + \sum_{j=1}^{m} c_j \delta_{t_j}, \\ G'(t) = bI(t)G(t) + aG(t), \\ I(0) = 0, \quad G(0) = G_0. \end{cases} \tag{2.41}$$

It is also important to investigate both optimal control problems under the control constraints

$$c_j \geq 0, \quad j \in \{1, 2, \ldots, m\}.$$

A better way to control the glucose level is, however, to act on insulin concentration (by injections) as well as on glucose concentration (by the food from usual meals).

2.5 Working examples

2.5.1 HIV treatment

We consider here a mathematical model that describes the interaction of the immune system with the HIV (human immunodeficiency virus) proposed in [KLS97]. Next we propose two optimal control problems based on chemotherapy which affects either the viral infectivity or the viral productivity.

The immune system is modeled in terms of the population of $CD4^+$ T cells (see [PKD93], [HNP95], and [PN02]). Let

$T(t)$ denote the concentration of uninfected $CD4^+$ T cells.
$T_i(t)$ denote the concentration of infected $CD4^+$ T cells.
$V(t)$ denote the concentration of free infectious virus particles

at moment t. The dynamics of the system is modeled by the following initial-value problem.

$$\begin{cases} T'(t) = \dfrac{s}{1+V(t)} - \mu_1 T(t) + rT(t)\left(1 - \dfrac{T(t)+T_i(t)}{T_{max}}\right) - k_1 V(t)T(t), \\[2mm] T_i'(t) = k_1 V(t)T(t) - \mu_2 T_i(t), \\[2mm] V'(t) = -k_1 V(t)T(t) - \mu_3 V(t) + N\mu_2 T_i(t), \\[2mm] T(0) = T_0, \quad T_i(0) = T_{i0}, \quad V(0) = V_0, \end{cases} \qquad (2.42)$$

$t \in (0, L)$ $(L > 0)$, where $s, k_1, r, N, \mu_1, \mu_2, \mu_3, T_{max}$ are positive constants and $T_0, T_{i0}, V_0 \geq 0$ are the initial concentrations of $CD4^+$ T cells, infected $CD4^+$ T cells, and free infectious virus particles, respectively.

The term $s/(1+V)$ represents a source term; the dependence upon the viral concentration V models the fact that infection of precursors of T cells may occur, thus reducing the production of the uninfected T cells.

The term $-k_1 VT$ in the first equation in (2.42) together with $+k_1 VT$ in the second equation in (2.42) models is the infection of T cells due to the viral concentration V; the term $-k_1 VT$ in the third equation in (2.42) models the binding of viruses to uninfected T cells, thus leading to infection.

μ_1, μ_2, μ_3 denote natural decay rates.

The term $N\mu_2 T_i$ in the third equation in (2.42) models the production of viruses during the decay of infected T cells.

The term

$$r\left(1 - \frac{T(t) + T_i(t)}{T_{\max}}\right)$$

represents the production rate of T cells.

Chemotherapy by a drug may either:

- Affect the virus infectivity, so that the second equation in (2.42) is modified into the following (see [BKL97]),

$$T_i'(t) = u(t)k_1 V(t)T(t) - \mu_2 T_i(t), \quad t \in (0, L),$$

  $u(t)$ being the control variable, that is, the strength of the chemotherapy. The first and third equations should be modified accordingly.
- Or reduce the viral production, which is most applicable to drugs such as protease inhibitors (see [KLS97]), thus modifying the third equation in (2.42) into

$$V'(t) = -k_1 V(t)T(t) - \mu_3 V(t) + u(t)N\mu_2 T_i(t), \quad t \in (0, L).$$

  The second equation should be modified accordingly.
  In either case the cost functional to maximize is

$$\int_0^L [aT(t) - \frac{1}{2}(1 - u(t))^2]dt,$$

$(a > 0)$ subject to $u \in L^2(0, L)$, $0 \le u(t) \le 1$ a.e. $t \in (0, L)$, which means maximizing the number of uninfected T cells, while simultaneously minimizing the "cost" of the chemotherapy to the human body.

A greater or lower value for a corresponds to a lower or greater importance given to minimizing the "cost" of the chemotherapy to the human body.

We propose that the reader derive the first-order necessary conditions of optimality for both optimal control problems.

Hint. The first optimal control problem proposed here is a particular case of (P1) (Section 2.1), for

$$G(t, u, T, T_i, V) = aT - \frac{1}{2}(1 - u)^2, \quad \varphi(T, T_i, V) = 0,$$

$$f(t, u, T, T_i, V) = \begin{pmatrix} \frac{s}{1 + V} - \mu_1 T + rT(1 - \frac{T + T_i}{T_{\max}}) - k_1 uVT \\ k_1 uVT - \mu_2 T_i \\ -k_1 uVT - \mu_3 V + N\mu_2 T_i \end{pmatrix}$$

and

$$K = \{w \in L^2(0, L); \ 0 \le w(t) \le 1 \text{ a.e. } t \in (0, L)\}.$$

The second optimal control problem proposed here is a particular case of (P1), for the same G, φ, K (as for the previous proposed problem), and

$$f(t, u, T, T_i, V) = \begin{pmatrix} \dfrac{s}{1+V} - \mu_1 T + rT(1 - \dfrac{T+T_i}{T_{\max}}) - k_1 VT \\ k_1 VT - \mu_2 u T_i \\ -k_1 VT - \mu_3 V + uN\mu_2 T_i \end{pmatrix}.$$

2.5.2 The control of a SIR model

We describe here the dynamics of a disease (transmitted only by contact between infectious and susceptible individuals) in a biological population using the following standard SIR model with vital dynamics (see [Cap93]).

$$\begin{cases} S'(t) = mN - mS(t) - cS(t)I(t) - u(t)S(t), \\ I'(t) = -mI(t) + cS(t)I(t) - dI(t), \\ R'(t) = -mR(t) + u(t)S(t) + dI(t), \end{cases} \tag{2.43}$$

for $t \in (0, L)$, $L > 0$, together with the initial conditions

$$S(0) = S_0 > 0, \quad I(0) = I_0 > 0, \quad R(0) = R_0 \geq 0. \tag{2.44}$$

Here

$S(t)$ represents the density of susceptible individuals,
$I(t)$ represents the density of infectious individuals, and
$R(t)$ represents the density of recovered (and immune) individuals

at moment t. $N = S(t) + I(t) + R(t) = S_0 + I_0 + R_0 > 0$ is a constant that represents the density of total population which is assumed to be constant.

Here m, c, d are positive constants. The incidence of the disease is described by the term $cS(t)I(t)$. The constant d represents the rate at which the infectious individuals recover.

The control u represents the part of the susceptible population being vaccinated. The vaccinated individuals recover.

We propose that the reader investigate the following optimal control problem for the above-mentioned SIR model:

$$\text{Minimize} \int_0^L [I(t) + au(t)^2]dt,$$

($a > 0$) subject to $u \in L^2(0, L)$, $0 \leq u(t) \leq M$ ($M > 0$) a.e. $t \in (0, L)$, where (S, I, R) is the solution to (2.43) and (2.44).

This means we are interested in minimizing the infectious population while simultaneously minimizing the "cost" of vaccination. A greater or lower value for a means a greater or lower importance given to minimizing the cost of vaccination.

Derive the maximum principle.

Hint. This problem is a particular case of (P1) (Section 2.1), for $m = 1$, $N = 3$, $T := L$,

$$G(t, u, S, I, R) = I + au^2, \quad \varphi(S, I, R) = 0,$$

$$f(t, u, S, I, R) = \begin{pmatrix} mN - mS - cSI - uS \\ -mI + cSI - dI \\ -mR + uS + dI \end{pmatrix},$$

and

$$K = \{w \in L^2(0, L); \ 0 \le w(t) \le M \text{ a.e. } t \in (0, L)\}.$$

Another important optimal control problem related to the SIR model proposed to the reader is the following identification problem,

$$\text{Minimize} \int_0^L [I(t) - \tilde{I}(t)]^2 dt,$$

subject to $c \in [0, M]$ ($M > 0$), where $\tilde{I} \in C([0, L])$, $\tilde{I}(t) \ge 0$ for any $t \in [0, L]$ is a known function and (S, I, R) is the solution to

$$\begin{cases} S'(t) = mN - mS(t) - cS(t)I(t), & t \in (0, L) \\ I'(t) = -mI(t) + cS(t)I(t) - dI(t), & t \in (0, L) \\ R'(t) = -mR(t) + dI(t), & t \in (0, L) \\ S(0) = S_0, \ I(0) = I_0, \ R(0) = R_0. \end{cases}$$

Here m, d, S_0, I_0, R_0 are given constants. The meaning of this problem is the following one. Knowing the number of infectious individuals at any moment we wish to determine the infectivity rate c.

Bibliographical Notes and Remarks

There is an extensive mathematical literature devoted to optimal control theory. This domain developed enormously after the pioneering work of Pontryagin and his collaborators. One of the main purposes when investigating an optimal control problem is to derive first-order necessary conditions of optimality (Pontryagin's principle). Here is a list of important monographs devoted to this subject: [LM67], [Kno81], [Bar93], [Bar94], and [Son98]. More applied optimal control problems can be found only in a few monographs; see [Kno81], [Che86], [Bar94], [Ani00], and [Tre05]. For applications in biology, with a few MATLAB programs we cite [LW07].

Exercises

2.1. Derive the maximum principle for the following problem:

$$\text{Maximize}\{x^u(T) + \gamma y^u(T)\},$$

subject to $u \in L^2(0,T)$, $0 \le u(t) \le 1$ a.e. $t \in (0,T)$, where (x^u, y^u) is the solution to the predator–prey system:

$$\begin{cases} x'(t) = r_1 x(t) - \mu_1 u(t) x(t) y(t), & t \in (0,T) \\ y'(t) = -r_2 y(t) + \mu_2 u(t) x(t) y(t), & t \in (0,T) \\ x(0) = x_0, \ y(0) = y_0. \end{cases}$$

Hint. Proceed as in Section 2.3. This problem is a particular case of (P1) (Section 2.1), for $m = 1$, $N = 2$,

$$G(t,u,x,y) = 0, \quad \varphi(x,y) = x + \gamma y,$$

$$f(t,u,x,y) = \begin{pmatrix} r_1 x - \mu_1 u x y \\ -r_2 y + \mu_2 u x y \end{pmatrix},$$

and

$$K = \{w \in L^2(0,T); \ 0 \le w(t) \le 1 \text{ a.e. } t \in (0,T)\}.$$

2.2. Derive the maximum principle for the following problem,

$$\text{Maximize}\{x^u(T) + y^u(T)\},$$

subject to $u \in L^2(0,T)$, $0 \le u(t) \le 1$ a.e. $t \in (0,T)$, where (x^u, y^u) is the solution to the predator–prey system

$$\begin{cases} x'(t) = r_1 x(t) - k x(t)^2 - \mu_1 u(t) x(t) y(t), & t \in (0,T) \\ y'(t) = -r_2 y(t) + \mu_2 u(t) x(t) y(t), & t \in (0,T) \\ x(0) = x_0, \ y(0) = y_0. \end{cases}$$

Here $r_1, r_2, k, \mu_1, \mu_2$ are positive constants, and kx represents an additional mortality rate and is due to the overpopulation; kx^2 is a logistic term for the prey population.

Hint. Proceed as in Section 2.1. This problem is a particular case of (P1) (Section 2.1).

2.3. Obtain the maximum principle for the following optimal harvesting problem:

$$\text{Maximize} \int_0^T u(t) x^u(t) dt,$$

subject to $u \in L^2(0,T)$, $0 \le u(t) \le M$ ($M > 0$) a.e. $t \in (0,T)$, where x^u is the solution to the following Malthusian model of population dynamics,

$$\begin{cases} x'(t) = r(t)x(t) - u(t)x(t), & t \in (0,T) \\ x(0) = x_0 > 0. \end{cases}$$

Here $x^u(t)$ represents the density of individuals of a population species at time t, $r \in C([0,T])$ gives the growth rate, and $u(t)$ is the harvesting effort (a control) and plays the role of an additional mortality rate. $\int_0^T u(t)x^u(t)dt$ represents the total harvested population on the time interval $[0,T]$.

Hint. Let u^* be an optimal control. Here are the first-order necessary optimality conditions:

$$\begin{cases} p'(t) = -r(t)p(t) + u^*(t)(1 + p(t)), & t \in (0,T) \\ p(T) = 0, \end{cases}$$

$$u^*(t) = \begin{cases} 0 & \text{if } 1 + p(t) < 0 \\ M & \text{if } 1 + p(t) > 0. \end{cases}$$

2.4. Obtain the maximum principle for the following optimal harvesting problem:

$$\text{Maximize} \int_0^T u(t)x^u(t)dt,$$

subject to $u \in L^2(0,T)$, $0 \le u(t) \le M$ ($M > 0$) a.e. $t \in (0,T)$, where x^u is the solution to the following logistic model of population dynamics,

$$\begin{cases} x'(t) = rx(t) - kx(t)^2 - u(t)x(t), & t \in (0,T) \\ x(0) = x_0 > 0. \end{cases}$$

Here r, k, x_0 are positive constants.

2.5. Derive the optimality conditions for the following problem,

$$\text{Maximize} \int_0^T u(t)x^u(t)dt - c \int_0^T u(t)^2 dt,$$

subject to $u \in L^2(0,T)$, $0 \le u(t) \le M$ ($M > 0$) a.e. $t \in (0,T)$, where x^u is the solution to the following logistic model of population dynamics,

$$\begin{cases} x'(t) = rx(t) - kx(t)^2 - u(t)x(t), & t \in (0,T) \\ x(0) = x_0 > 0. \end{cases}$$

Here c, r, k, x_0 are positive constants. This problem seeks to maximize the harvest while minimizing effort.

3

Optimal control of ordinary differential systems. Gradient methods

This chapter is devoted to approximation methods, mainly of gradient type, for optimal control problems governed by ordinary differential equations. The main goal is to build corresponding MATLAB® programs. The calculation of the gradient of the cost functional allows us to develop gradient-type algorithms. We deal with minimization/maximization problems. As we show, the general principle of a gradient method is the same for both types of problems.

3.1 A gradient method

This section represents an introduction to the numerical approximation of control problems by gradient methods. To develop the algorithm we take as our example the abstract optimal control problem from Section 2.1. A gradient method can be used for minimization/maximization problems, the principle being the same in both situations. Such a method is an iterative one which makes a local search at each iteration to improve the value of the cost functional (to increase it for maximization and to decrease it for minimization). If the current control is $u_k = u^{(k)}$, then we successively compute x^{u_k} (the solution of the state equation with input u_k), p^{u_k} (the solution of the adjoint equation with inputs u_k, x^{u_k}), and the gradient $\Phi_u(u_k)$, also denoted by $\nabla_u \Phi(u_k)$. A formula that also contains p^{u_k} (but not x^{u_k}) is obtained for the gradient. This approach, called the elimination of the state, was introduced by J. Céa in [Cea78]. As an example we consider the control problem from Section 2.1 in the case $\varphi = 0$.

$$\text{Maximize } \Phi(u) = \mathcal{L}(u, x^u) = \int_0^T G(t, u(t), x^u(t))dt, \qquad \text{(P1}')$$

subject to $u \in K \subset U = L^2(0, T; I\!\!R^m)$ $(T > 0)$, where x^u is the solution to

$$\begin{cases} x'(t) = f(t, u(t), x(t)), & t \in (0, T) \\ x(0) = x_0. \end{cases} \qquad (3.1)$$

S. Aniţa et al. *An Introduction to Optimal Control Problems in Life Sciences and Economics*, Modeling and Simulation in Science, Engineering and Technology, DOI 10.1007/978-0-8176-8098-5_3, © Springer Science+Business Media, LLC 2011

Here

$$G : [0, T] \times I\!R^m \times I\!R^N \to I\!R,$$
$$f : [0, T] \times I\!R^m \times I\!R^N \to I\!R^N,$$

$x_0 \in I\!R^N$, and $K \subset U$ is a closed convex subset.

Because the cost functionals of the models in this chapter do not involve the final value $x(T)$ we have removed the term $\varphi(x^u(T))$ present in Section 2.1. However, the approach is similar even if $\varphi(x^u(T))$ occurs in the cost functional. Problem (P1') may also be reformulated as a minimization problem

$$\text{Minimize } \Psi(u) = -\mathcal{L}(u, x^u),$$

subject to $u \in K \subset U$, where $\Psi(u) = -\Phi(u)$.

Assume that (u^*, x^{u^*}) is an optimal pair for Problem (P1'). For the time being we assume that G and f are sufficiently smooth and that all the following operations are allowed.

We assume that the function $u \mapsto x^u$ is everywhere Gâteaux differentiable. We denote by dx^u this differential. Consider the arbitrary but fixed elements $u, v \in U$ and

$$V = \{v \in U; \ u + \varepsilon v \in K \text{ for any } \varepsilon > 0 \text{ sufficiently small}\}.$$

For any $v \in V$, we define $z = dx^u(v)$; z is the solution to

$$\begin{cases} z'(t) = f_u(t, u(t), x^u(t))v(t) + f_x(t, u(t), x^u(t))z(t), & t \in (0, T) \\ z(0) = 0. \end{cases} \tag{3.2}$$

For an arbitrary but fixed $v \in V$ we have

$$(v, \Phi_u(u)) = \lim_{\varepsilon \to 0+} \frac{1}{\varepsilon} [\Phi(u + \varepsilon v) - \Phi(u)],$$

where $(\cdot, \cdot)$ is the inner product on U, and consequently

$$(v, \Phi_u(u)) = \int_0^T [v(t) \cdot G_u(t, u(t), x^u(t)) + z(t) \cdot G_x(t, u(t), x^u(t))]dt. \tag{3.3}$$

Let p^u be the solution to the adjoint problem:

$$\begin{cases} p'(t) = -f_x^*(t, u(t), x^u(t))p(t) - G_x(t, u(t), x^u(t)), & t \in (0, T) \\ p(T) = 0. \end{cases} \tag{3.4}$$

We multiply the equation in (3.2) by p^u and we integrate by parts on $[0, T]$. If we take into account (3.4) we get as in Section 2.1 that

$$\int_0^T z(t) \cdot G_x(t, u(t), x^u(t))dt = \int_0^T v(t) \cdot f_u^*(t, u(t), x^u(t))p^u(t)dt. \quad (3.5)$$

By (3.5) and (3.3) we finally obtain

$$\Phi_u(u) = G_u(., u, x^u) + f_u^*(., u, x^u)p^u. \quad (3.6)$$

By using (3.1), (3.4), and (3.6) we derive an iterative method to improve the value of the cost functional at each step (or to approximate the optimal control u^*). We use the gradient of the cost functional Φ to get an increase of the cost functional $\Phi(u)$ at each iteration. We have to apply a projected gradient method in order to handle the control constraint $u \in K$. We write the corresponding version of Uzawa's algorithm for Problem (P1') (e.g., [AN03, Section 2.5]):

S0: Choose $u^{(0)} \in K$;
        Set $k := 0$.
S1: Compute $x^{(k)}$ the solution to (3.1) corresponding to $u := u^{(k)}$:

$$\begin{cases} x'(t) = f(t, u^{(k)}(t), x(t)), & t \in (0, T) \\ x(0) = x_0. \end{cases}$$

S2: Compute $p^{(k)}$ the solution to (3.4) for $u := u^{(k)}$ and $x := x^{(k)}$:

$$\begin{cases} p'(t) = -f_x^*(t, u^{(k)}(t), x^{(k)}(t))p(t) - G_x(t, u^{(k)}(t), x^{(k)}(t)), & t \in (0, T) \\ p(T) = 0. \end{cases}$$

S3: Compute the gradient $w^{(k)}$ using formula (3.6):

$$w^{(k)} := \Phi_u(u^{(k)}) = G_u(\cdot, u^{(k)}, x^{(k)}) + f_u^*(\cdot, u^{(k)}, x^{(k)})p^{(k)}.$$

S4: Compute the steplength $\rho_k \geq 0$ such that

$$\Phi(P_K(u^{(k)} + \rho_k w^{(k)})) = \max_{\rho \geq 0} \Phi(P_K(u^{(k)} + \rho w^{(k)})).$$

S5: $u^{(k+1)} := P_K(u^{(k)} + \rho_k w^{(k)})$.
S6: **(The stopping criterion)**
        If $\|u^{(k+1)} - u^{(k)}\| < \varepsilon$
            then STOP ($u^{(k+1)}$ is the approximating control)
            else $k := k + 1$; go to S1.

We point out that for a minimization gradient method the general form of the algorithm is the same with only two "small" changes:

- In Step S3 we take $w^{(k)} := -\Phi_u(u^{(k)})$ to obtain a descent direction.
- In Step S4 we take $\min_{\rho \geq 0}$ instead of $\max_{\rho \geq 0}$.

In the current literature such an algorithm is known as the steepest descent method. Moreover "direction" means a vector, the sense being defined by $\rho > 0$. It is also possible to use in practice some other stopping criterion as we discuss for concrete examples.

Let us mention that P_K above is the projection operator on the convex set K; that is, $P_K : U \to K$ is defined by

$$\|P_K(u) - u\| \leq \|w - u\| \quad \text{for any } w \in K,$$

where $\|\cdot\|$ denotes the norm of U. $\varepsilon > 0$ in step S6 is a prescribed precision.

If the control is not restricted (i.e., $K = U$), then $P_K = P_U = I_U$ and S4 becomes the following.

Compute the steplength $\rho_k \geq 0$ such that

$$\Phi(u^{(k)} + \rho_k w^{(k)}) = \max_{\rho \geq 0}\{\Phi(u^{(k)} + \rho w^{(k)})\}.$$

A tedious practical problem in the above algorithm is to compute the steplength ρ_k from S4. We need a robust and efficient procedure to do it. The problem to be solved many times in S4 is the following one: given some value $\rho > 0$, compute $\Phi(P_K(u^{(k)} + \rho w^{(k)}))$. Such trials are necessary to find a proper steplength ρ_k. To do this we have to consider the following substeps.

- Compute $u := u^{(k)} + \rho w^{(k)}$.
- Compute $\bar{u} := P_K(u)$.
- Compute $\bar{x}$, the solution to (3.1) corresponding to $u := \bar{u}$.
- Compute the corresponding value of the cost functional $\Phi(\bar{u})$.

Clearly this subproblem requires a large amount of computation. For details about descent methods and efficient algorithms to compute the steplength ρ_k we refer to [AN03, Section 2.3], and to [GS81, Chapter 2].

For the particular case when $K = U$ the following simplified algorithm can be used.

The forward–backward sweep method

NS′0: Choose $u^{(0)} \in U$.
Set $k := 0$.
NS′1: Compute $x^{(k)}$ the solution to (3.1) corresponding to $u := u^{(k)}$:

$$\begin{cases} x'(t) = f(t, u^{(k)}(t), x(t)), & t \in (0, T) \\ x(0) = x_0. \end{cases}$$

NS′2: Compute $p^{(k)}$ the solution to (3.4) corresponding to $u := u^{(k)}$:

$$\begin{cases} p'(t) = -f_x^*(t, u^{(k)}(t), x^{(k)}(t))p(t) - G_x(t, u^{(k)}(t), x^{(k)}(t)), & t \in (0, T) \\ p(T) = 0. \end{cases}$$

NS′3: Compute $u^{(k+1)}$ the solution to the equation:

$$G_u(t, u(t), x^{(k)}(t)) + f_u^*(t, u(t), x^{(k)}(t))p^{(k)} = 0;$$

NS′4: (The stopping criterion)
   If $\|u^{(k+1)} - u^{(k)}\| < \varepsilon$
   then STOP ($u^{(k+1)}$ is the approximating control)
   else $k := k + 1$; go to NS′1.

If constraints on the control occur then a descent forward–backward sweep method can be used.

We now present a short outline of the theory of the steepest descent method for minimization problems and the relationship with maximization problems. We consider the problem of minimizing $\Psi : U \mapsto \mathbb{R}$ over the real Hilbert space U. We assume that U is identified with its own dual. Here we use the traditional notation $\nabla\Psi$ for Ψ_u because Ψ depends only on u. A class of iterative minimization methods is defined by

$$u^{(k+1)} = u^{(k)} + \rho_k w^{(k)}, \tag{3.7}$$

where $w^{(k)} \in U$ is a *search direction* and $\rho_k \in \mathbb{R}^+$ is the *steplength*. We say that $w \in U$ is a *descent direction* for Ψ at $u \in U$ if there exists $\rho_0 > 0$ such that

$$\Psi(u + \rho w) < \Psi(u) \text{ for any } \rho \in (0, \rho_0]. \tag{3.8}$$

Assume now that Ψ has a continuous gradient on U. Then a descent direction w at $u \in U$ occurs when

$$(w, \nabla\Psi(u)) < 0 \tag{3.9}$$

where $(\cdot, \cdot)$ is the inner product of U, and $\|\cdot\|$ is the corresponding norm.

Moreover the steepest descent direction is obtained as follows. If $\nabla\Psi(u) \neq 0$, then we apply Taylor's formula and we get

$$\Psi(u + \rho w) = \Psi(u) + \rho(w, \nabla\Psi(u)) + o(\rho).$$

If w is a descent direction at u then (3.9) is satisfied and therefore the steepest descent is obtained for $(w, \nabla\Psi(u))$ minimal. Hence we search for

$$\min \{(w, \nabla\Psi(u)) \; ; \; \|w\| = 1\}. \tag{3.10}$$

By Schwarz's inequality we readily get

$$|(w, \nabla\Psi(u))| \leq \|\nabla\Psi(u)\|$$

and therefore

$$-\|\nabla\Psi(u)\| \leq (w, \nabla\Psi(u)) \text{ if } \|w\| = 1.$$

It is clear now that $w = -\nabla\Psi(u)/\|\nabla\Psi(u)\|$ gives the minimum value for (3.10). The steepest descent direction at u is $-\nabla\Psi(u)$.

Assume now we want to maximize $\Psi(u)$ for $u \in U$. A direction w that ensures the local increase of Ψ at u is defined similarly to a descent direction and is characterized by

$$(w, \nabla\Psi(u)) > 0. \tag{3.11}$$

Also using Taylor's formula it follows that we should search for

$$\max\{(w, \nabla\Psi(u)) \; ; \; \|w\| = 1\} \tag{3.12}$$

in order to obtain a maximum increase of Ψ at u. By Schwarz's inequality we readily get

$$(w, \nabla\Psi(u)) \le \|\nabla\Psi(u)\| \quad \text{if } \|w\| = 1.$$

It is clear now that $w = \nabla\Psi(u)/\|\nabla\Psi(u)\|$ gives the maximum value for (3.12). Therefore the local maximum increase of Ψ at u is obtained for the direction $\nabla\Psi(u)$.

Therefore the main difference between minimization and maximization problems consists in the sign considered for the gradient. But the theory of descent directions and of the convergence of such methods is more complicated, a very important role being played by the sequence $\{\rho_k\}$ of steplengths (e.g., [AN03, Chapter 2], [GS81, Chapters 7 and 8]). There one can also find more about constrained controls and projected gradient methods. Such problems are discussed for concrete examples in the next sections.

We pass now to the control problems. Consider

$$\text{Minimize } \Psi(u),$$

subject to $u \in K \subset U$, where K is a closed convex subset. If the functional Ψ is strictly convex and smooth enough (has continuous gradient) and if K is bounded or Ψ is coercive on K ($\lim_{\|u\|\to\infty} \Psi(u) = +\infty$) the convergence of the projected gradient method to the unique optimal control has been proved. If the control is not constrained ($u \in U$), $u \in U$, then the convergence is valid for the gradient method (e.g., [AN03, Section 2.1]). Otherwise the convergence theory is more complicated (e.g., [Pol71], [Cea78], [GS81], and [AN03]). Generally speaking we have an algorithm that obtains a decrease/increase by iteration for the cost functional.

The forward–backward sweep method is a direct consequence of Pontryagin's principle. If we denote $u^{(k+1)} = \Gamma u^{(k)}$, where Γ is the operator corresponding to this algorithm, then the iterative method from Banach's fixed point theorem can be used to obtain the convergence of the method (if Γ is a contraction).

3.2 A tutorial example for the gradient method

We now present the application of a gradient method, the mathematical calculation of the gradient, and the corresponding program. We consider a simple tutorial example that allows us, however, to illustrate an efficient method to approximate optimal control and to compare the approximated control with the calculated one.

In a (bio)chemical reaction a component is added at a constant rate over a time interval $[0, L]$ $(L > 0)$ ([Kno81, Chapter III, Section 1]). Let $x(t)$ be the deviation of the pH value from the desired one at time t. We have to control the pH value because the quality of the product depends on it. The control is made by the strength $u(t)$ of the controlling ingredient. The dynamics of x is described by the following initial-value problem:

$$\begin{cases} x'(t) = ax(t) + \beta u(t), \ t \in (0, L) \\ x(0) = x_0, \end{cases} \tag{3.13}$$

where α and β are known positive constants and x_0 is the initial deviation of the value of the pH. We suppose that the decrease in yield of the deviation of the pH due to changes in the pH is $\int_0^L x(t)^2 dt$. We also suppose that the rate of cost of keeping the strength u is proportional to u^2. Therefore this (bio)chemical model leads to the following optimal control problem,

$$\text{Minimize } \frac{1}{2} \int_0^L [ax^u(t)^2 + u(t)^2] dt, \tag{TP}$$

subject to $u \in U = L^2(0, L)$, where x^u is the solution to (3.13), and a is a positive constant. There are no constraints on controls.

Problem (TP) is of course a particular case of (P1) (in Section 2.1), for $m = 1$, $N = 1$, $T = L$,

$$G(t, u, x) = -\frac{1}{2}(ax^2 + u^2),$$
$$\varphi(x) = 0,$$
$$f(t, u, x) = \alpha x + \beta u.$$

Here

$$\mathcal{L}(u, x) = -\frac{1}{2} \int_0^L [ax(t)^2 + u(t)^2] dt,$$

$$\Phi(u) = -\frac{1}{2} \int_0^L [ax^u(t)^2 + u(t)^2] dt,$$

and problem (TP) may be reformulated as

$$\text{Minimize}\{-\Phi(u)\}, \text{ subject to } u \in U.$$

To make the explanation of the method easier we denote by $\Psi(u)$ the cost functional of the original problem, that is,

$$\Psi(u) = \frac{1}{2} \int_0^L [ax^u(t)^2 + u(t)^2]dt,$$

and our problem becomes

Minimize $\Psi(u)$,

subject to $u \in U$. First of all we calculate the gradient of the cost functional. As already noticed this leads to the necessary conditions of optimality. From a computational point of view this means to calculate the gradient of the cost functional via the adjoint state (by elimination of the state).

Consider two arbitrary but fixed elements $u, v \in U = L^2(0, L)$. Let z be the solution to the following IVP (initial-value problem),

$$\begin{cases} z'(t) = \alpha z(t) + \beta v(t), \ t \in (0, L) \\ z(0) = 0. \end{cases} \qquad (3.14)$$

We now introduce the adjoint problem (the adjoint state is p^u):

$$\begin{cases} p'(t) = -\alpha p(t) + ax^u(t), \ t \in (0, L) \\ p(L) = 0. \end{cases} \qquad (3.15)$$

For any $\varepsilon \in \mathbb{R}^*$ we have

$$\Psi(u + \varepsilon v) - \Psi(u) = \frac{1}{2} \int_0^L [ax^{u+\varepsilon v}(t)^2 + (u(t) + \varepsilon v(t))^2 - ax^u(t)^2 - u(t)^2]dt.$$

Because problem (3.13) is linear, it follows immediately that

$$x^{u+\varepsilon v} = x^u + \varepsilon z,$$

and consequently

$$\frac{1}{\varepsilon}[\Psi(u + \varepsilon v) - \Psi(u)] = \frac{1}{2} \int_0^L [2ax^u(t)z(t) + \varepsilon az(t)^2 + 2u(t)v(t) + \varepsilon v(t)^2]dt.$$

We let $\varepsilon \to 0$ and we get

$$(v, \Psi_u(u)) = \int_0^L [ax^u(t)z(t) + u(t)v(t)]dt, \qquad (3.16)$$

where $(\cdot, \cdot)$ is the usual inner product on $L^2(0, L)$, defined by

$$(g_1, g_2) = \int_0^L g_1(t)g_2(t)dt, \quad g_1, g_2 \in L^2(0, L).$$

Next we wish to eliminate x^u from (3.16). By multiplying the differential equation in (3.14) by p^u, and by integrating on $[0, L]$ we get

$$\int_0^L z'(t)p^u(t)dt = \int_0^L [\alpha z(t) + \beta v(t)]p^u(t)dt,$$

and consequently, because $z(0) = p(L) = 0$,

$$-\int_0^L z(t)(p^u)'(t)dt = \int_0^L [\alpha z(t) + \beta v(t)]p^u(t)dt.$$

By (3.14) and (3.15) we obtain that

$$-\int_0^L z(t)[-\alpha p^u(t) + ax^u(t)]dt = \int_0^L [\alpha z(t) + \beta v(t)]p^u(t)dt,$$

and then

$$\int_0^L ax^u(t)z(t)dt = -\int_0^L \beta v(t)p^u(t)dt. \tag{3.17}$$

By (3.17) and (3.16) we infer that

$$(v, \Psi_u(u)) = \int_0^L v(t)[u(t) - \beta p^u(t)]dt.$$

This is true for any $v \in L^2(0, L)$, therefore we may conclude that

$$\Psi_u(u) = u - \beta p^u. \tag{3.18}$$

If u^* is the optimal control for (TP), then

$$\Psi_u(u^*) = u^* - \beta p^{u^*},$$

where p is the solution to (3.15) corresponding to $u := u^*$.

We are able now to write a gradient algorithm using $-\Psi_u(u)$ as the descent direction for the current control u.

A gradient algorithm for problem (TP)

S0: Choose $u^{(0)} \in U$;
   Set k:=0.
S1: Compute $x^{(k)}$, the solution to (3.13) with the input $u^{(k)}$:

$$\begin{cases} x'(t) = \alpha x(t) + \beta u^{(k)}(t), & t \in (0, L) \\ x(0) = x_0. \end{cases}$$

S2: Compute $p^{(k)}$, the solution to (3.15) with the input $x^{(k)}$:

$$\begin{cases} p'(t) = -\alpha p(t) + ax^{(k)}(t), & t \in (0, L) \\ p(L) = 0. \end{cases}$$

S3: Compute the gradient $w^{(k)}$ using formula (3.18):

$$w^{(k)} := \Psi_u(u^{(k)}) = u^{(k)} - \beta p^{(k)}.$$

S4: Compute the steplength ρ_k such that

$$\Psi(u^{(k)} - \rho_k w^{(k)}) = \min_{\rho \geq 0} \{\Psi(u^{(k)} - \rho w^{(k)})\}.$$

S5: Compute the new control

$$u^{(k+1)} := u^{(k)} - \rho_k w^{(k)}.$$

S6: (The stopping criterion)
   If $\|u^{(k+1)} - u^{(k)}\| < \varepsilon$
       then STOP ($u^{(k+1)}$ is the approximating control)
       else $k := k + 1$; go to S1.

The norm used in Step S6 is a discrete one and $\varepsilon > 0$ is a prescribed precision. Moreover we show later that other stopping criteria can be used in practice.

Now we make an important comment for practical purposes. In many problems there are control restrictions; that is, we have $u \in K \subset U$ with K a closed convex subset. In such a case a projected gradient algorithm should be used (e.g., [AN03, Section 2.5]). For instance, for Uzawa's algorithm, the formula in Step S4 above becomes

$$\Psi(P_K(u^{(k)} - \rho_k w^{(k)})) = \min_{\rho \geq 0} \{\Psi(P_K(u^{(k)} - \rho w^{(k)}))\},$$

where $P_K : U \to K$ is the corresponding projection operator (e.g., [AN03, Section 2.4]). Of course the formula in Step S5 becomes

$$u^{(k+1)} := P_K(u^{(k)} - \rho_k w^{(k)}).$$

To apply such an algorithm we need to get a computable formula for P_K that can be implemented in a computer program. For instance, a usual restriction for the control is

$$|u(t)| \leq M \text{ a.e. } t \in (0, L)$$

($K = \{w \in L^2(0, L); |w(t)| \leq M \text{ a.e. } t \in (0, L)\}$). In such a case we have $P_K : L^2(0, L) \to K$,

$$P_K(u)(t) = Proj(u(t)) \text{ a.e. } t \in (0, L),$$

where

$$Proj(w) = \begin{cases} w & \text{if } -M \leq w \leq M \\ -M & \text{if } w < -M \\ M & \text{if } w > M. \end{cases}$$

A function used later follows from the above formula.

```
function y = Proj(u)
global M
y = u ;
if u < −M
    y = −M ;
end
if u > M
    y = M ;
end
```

We are now going to build the program for the gradient algorithm described above. For ease of understanding we give successive parts of the program explaining the implementation of every step of the algorithm. The entire program can be obtained by simple concatenation of all parts.

We point out first that we use a time grid with equidistant knots as follows,

$$t_i = (i - 1)h, \quad i = 1, \ldots, n,$$

where

$$h = \frac{L}{n - 1},$$

and the vectors

$$uold \text{ for } u^{(k)}, \quad unew \text{ for } u^{(k+1)}.$$

We start with *Part 1* which contains the usual beginning statements, input of data, and step S0 of the algorithm. Let us point out that the starting control $u^{(0)}(t)$ is constant, and that the meaning of some input variables is explained later.

```
% file CONT0.m
% Optimal control for the pH in a (bio)chemical reaction
% gradient algorithm with Armijo method for the steplength ρ
% ============
%      PART 1
% ============
clear
global alf bet ind
global a
global h
global uold
global u
global x
L = input('final time : ') ;
alf = input('alpha : ') ; % α
bet = input('beta : ') ; % β
```

```
a = input('a : ') ;
x0 = input('x(0) : ') ;
h = input('h : ') ; % time grid step
t = 0:h:L ; % time grid
t = t' ; % change to column vector
n = length(t) ;
x = zeros(n,1) ; % state vector
x1 = zeros(n,1) ; % state vector to be used for ρ loop (Armijo method)
p = zeros(n,1) ; % adjoint state vector
uold = zeros(n,1) ; % control vector corresponding to u^(k)
unew = zeros(n,1) ; % control vector corresponding to u^(k+1)
u = zeros(n,1) ; % control vector to be used for ρ loop (Armijo method)
% S0 : control initialization
u0 = input('u0 : ') ;
for i = 1:n
    uold(i) = u0 ;
end
disp('enter control data') ;
eps = input('precision : ') ; % precision ε for the gradient algorithm
maxit = input('maxiter : ') ; % max. no. iterations – gradient algorithm
disp('RO data') ;
roin = input('RO : ') ; % initial value for gradient steplength ρ
bro = input('b for RO : ') ; % b parameter for ρ loop
eps1 = input('precision for RO : ') ; % precision ε1 for steplength ρ
maxro = input('max for RO : ') ; % max. no. of iterations – ρ loop
ro = roin ; % initialization of steplength ρ
flag1 = 0 ; % convergence indicator for gradient algorithm
ii = 0 ; % index for gradient values
```

The values of $flag1$ are:

0 – No convergence obtained for the gradient algorithm.
1 – Convergence obtained for the gradient algorithm.

The variable ii keeps the index of vector $grad$ which stores the values of the gradient $\Psi_u(u^{(k)})$ for successive iterations. This vector is created component after component at each iteration. We continue the program by

```
% ===========
%      PART 2
% ===========
% gradient loop starts
for iter = 1:maxit
    iter
    % S1 : solve the state equation (3.13) by Runge–Kutta method
    x(1) = x0 ; % the initial condition
    for i = 1:n – 1
```

```
    ind = i ; % ind is a global index variable used by the rhs of (3.13)
    x(i + 1) = RK41(x(i),t(i)) ;
end
disp('SE solved') ;
if iter == 1
    Q = a*x.∧2 + uold.∧2 ;
    temp = trapz(t,Q) ;
    cvold = temp/2
    cost(1) = cvold ;    % store the value of the cost functional
    jj = 1 ;
end
```

Problem (3.13) is solved by the standard Runge–Kutta method of order 4
(see Appendix A.4). We do not use an *ode* function because it is more complicated to provide the control u. MATLAB asks for a function, but we have
a vector and for every given time moment t we need only one value of the
vector. We therefore decided to provide it using the global variable *ind* and
the corresponding value *uold(ind)*. The function *RK41.m* is exactly *RK4.m*
from Chapter 1. For the sake of clarity we give it once again.

```
function yout = RK41(x,t)
global h
tm = t + h/2 ;
k1 = h * F1(t,x) ;
k2 = h * F1(tm,x + k1/2) ;
k3 = h * F1(tm,x + k2/2) ;
k4 = h * F1(t + h,x + k3) ;
yout = x + (k1 + k4 + 2.0*(k2 + k3))/6.0 ;
```

The corresponding function *F1.m* is

```
function yout = F1(t,x)
global alf bet ind
global uold
yout = alf*x + bet*uold(ind) ;
```

For *iter* $= 1$ we compute the value of the cost functional, using the function
trapz, and we store it in *cvold* and *cost(1)*. Here jj is the index of vector *cost*
which stores the values of the cost functional $\Psi(u^{(k)})$ for successive iterations.
This vector is built component after component at each iteration.

We now handle Step S2 from the algorithm. For the adjoint state equation
(3.15) we still use the Runge–Kutta standard method of order 4. The only
difference is we have to proceed from L descending to 0. Then the gradient is
computed.

```
% ============
%     PART 3
% ============
    % S2 : solve adjoint equation (3.15)
    p(n) = 0 ;
    for j = 1:n − 1
        i = n − j ;
        ip1 = i + 1 ;
        ind = ip1 ;
        p(i) = RK43(p(ip1),t(ip1)) ;
    end
    disp('AE solved') ;
    % S3 : compute the gradient
    w = uold − bet*p ;
    disp('GRADIENT COMPUTED') ;
```

The file function *RK43.m* for the descending Runge–Kutta method is the following.

```
function yout = RK43(q,t)
global h
h1 = −h ;
tm = t + h1/2 ;
k1 = h1 * F3(t,q) ;
k2 = h1 * F3(tm,q + k1/2) ;
k3 = h1 * F3(tm,q + k2/2) ;
k4 = h1 * F3(t + h1,q + k3) ;
yout = q + (k1 + k4 + 2.0*(k2 + k3))/6.0 ;
```

The function file *F3.m* follows.

```
function yout = F3(t,p)
global alf bet ind
global a
global x
yout = a*x(ind) − alf*p ;
```

Before going to S4 we discuss the stopping criteria of the algorithm. One can be found in Step S6, but as already asserted we use more such criteria; that is,

- **SC1** : $\|\Psi_u(u^{(k)})\| < \varepsilon$.
- **SC2** : $|\Psi(u^{(k+1)}) − \Psi(u^{(k)})| < \varepsilon$.
- **SC3** : $\|u^{(k+1)} − u^{(k)}\| < \varepsilon$.
- **SC4** : $\rho_k < \varepsilon_1$.

Let us remark that SC1 and SC3 are equivalent because

$$\|u^{(k+1)} − u^{(k)}\| = \rho_k \|\Psi_u(u^{(k)})\|.$$

The test SC4 says that the gradient steplength ρ is very small which means that numerically no descent can be obtained along the direction $-w^{(k)}$, where $w^{(k)} = \Psi_u(u^{(k)})$. We can therefore assert that SC4 is numerically equivalent to SC2. The policy for our program is to stop when one convergence criterion is satisfied. Inasmuch as the gradient $\Psi_u(u^{(k)})$ was just computed we apply SC1 and we continue the program by

```
% ============
%      PART 4
% ============
    normg = sqrt(sum(w.^2))    % compute the gradient norm
    ii = ii + 1 ;
    grad(ii) = normg ;
    % verify SC1
    if normg < eps
        disp('CONVERGENCE by GRADIENT')
        flag1 = 1 ;
        break
    end
```

If the stopping criterion is satisfied (i.e., *normg* < *eps*), we give to the convergence indicator *flag1* the corresponding value 1 and we use the *break* statement which moves out the gradient loop started in PART 2 by the statement *for iter = 1:maxit*. We proceed now to Step S5 of the gradient algorithm, a more difficult one, which requires the computation of the gradient steplength ρ_k. The formula in Step S5 is useless in practice because usually the minimum there cannot be calculated. We approximate it by a robust and efficient algorithm, namely the Armijo method (for more details see [A66], [AN03, Section 2.3] and [Pol71, Appendix C.2]). We have already computed $\Psi(u^{(k)})$. The last known value for the steplength ρ is ρ_{k-1}. We also know $b \in (0.5, 0.8)$ the parameter to fit ρ (by the *input* of the variable *bro*). Let us denote for what follows $\Psi(u) = \psi(u, x^u)$ in order to emphasize the current state x also. Here

$$\psi(u, x) = \frac{1}{2} \int_0^L [ax(t)^2 + u(t)^2] dt.$$

Armijo method for the gradient steplength

A0: Set $\overline{\rho} := \rho_{k-1}$.
A1: $u := u^{(k)} - \overline{\rho} w^{(k)}$.
A2: Compute x_1 the solution of the state problem with input u:

$$\begin{cases} x_1'(t) = \alpha x(t) + \beta u(t), & t \in (0, L) \\ x_1(0) = x_0. \end{cases}$$

A3: Compute $\psi(u, x_1)$.

A4: (The stopping criterion)
If $\psi(u, x_1) \geq \psi(u^{(k)}, x^{(k)})$
then $\bar{\rho} := b\bar{\rho}$; go to A1
else $\rho_k := \bar{\rho}$; STOP algorithm.

The value ρ_k is used next in Step S5 of the descent algorithm and in the next gradient iteration as the starting value for $\bar{\rho}$.

The program variables corresponding to the mathematical ones in the above algorithm are:

- ρ_{k-1} = ro ;
- $\bar{\rho}$ = robar ;
- u = u ;
- x_1 = x1 ;
- $\psi(u^{(k)}, x^{(k)})$ = cvold ;
- $\psi(u, x_1)$ = cv ;
- b = bro ;
- ρ_k = ro (actualized later) .

The Armijo method loop is a *for* statement controlled by the variable *count* and by the convergence indicator *flag2*. The values of *flag2* are:

0 – No convergence obtained for Armijo algorithm.
1 – Convergence obtained for Armijo algorithm.

Therefore the next part of the program is as follows.

```
% ============
%      PART 5
% ============
   % S4 : FIT RO
   robar = ro ;
   flag2 = 0 ;
   flag3 = 0 ;
   % start loop to fit ro
   for count = 1:maxro
      count
      u = uold − robar*w ;
      % solve state equation for input u and get state x1
      x1(1) = x0 ;
      for j = 1:n − 1
         ind = j ;
         x1(j + 1) = RK42(x1(j),t(j)) ;
      end
      % test for SC4
      if robar < eps1
         flag3 = 1 ;
```

```
      break    % leave loop for count ...
    end
    % compute current cost value and test ρ̄
    Q1 = a*x1.∧2 + u.∧2 ;
    temp = trapz(t,Q1) ;
    cv = temp/2 ;
    if cv >= cvold    % no decrease of cost for minimization
        robar = bro * robar
    else
        flag2 = 1 ;
        break % leave loop for count ...
    end
  end % for count
```

The state equation (3.13) is solved for control u to get state x_1, the same way as before. The only difference is a little programming problem. We use as control u instead of $uold$. We therefore call $RK42$ instead of $RK41$. The function $RK42$ is a copy of $RK41$ where the call of $F1$ is replaced by the call of $F2$. The function $F2$ follows.

```
function yout = F2(t,x)
global alf bet ind
global u
yout = alf*x + bet*u(ind) ;
```

Next we take the conclusion for the above loop (with variable *count*). If $flag2 = 1$ then the Armijo algorithm is convergent and we get a new value for the cost functional (stored by *cvnew*). The loop can also end by criterion SC4; that is, $\bar{\rho} < \varepsilon_1$ with $flag3 = 1$. In such a case the current $u^{(k)}$ is numerically optimal and in the outer loop *for iter* ...the value of $flag3$ determines the program to stop. In the next sequence we also verify the stopping criteria SC2 and SC3. Otherwise the whole gradient algorithm fails because the Armijo procedure was not able to fit the steplength ρ_k. The corresponding part of the program follows.

```
% ============
%      PART 6
% ============
    if flag3 == 1
        disp('CONVERGENCE BY RO')
        break % leave loop for iter ... (SC4 satisfied)
    end
    if flag2 == 1
        cvnew = cv
        jj = jj + 1 ;
        cost(jj) = cvnew ;
        unew = u ;
```

```
        ro = robar ;
        % verify SC2
        if abs(cvnew − cvold) < eps
            disp('CONVERGENCE by COST')
            flag1 = 1 ;
            break % leave the loop for iter ... (SC2 satisfied)
        end
        % verify SC3
        d = uold − unew ;
        dif = sqrt(sum(d.∧2)) ;
        if dif < eps
            disp('CONVERGENCE by CONTROL')
            flag1 = 1 ;
            break % leave the loop for iter ... (SC3 satisfied)
        end
    else
        error('NO CONVERGENCE FOR RO') % STOP PROGRAM
    end
```

Next we close the loop of the gradient algorithm and we display results and print figures if convergence is obtained.

```
    % ===========
    %      PART 7
    % ===========
        % prepare a new iteration
        uold = unew ;
        cvold = cvnew ;
        % x = x1 ; to be introduced in the final version below
    end % for iter
    if (flag1 == 1) | (flag3 == 1) % | means logical or
        plot(t,unew,'*') ; grid
        xlabel('\bf t','FontSize',16)
        ylabel('\bf u(t)','FontSize',16)
        figure(2)
        plot(t,x1,'r*') ; grid
        xlabel('\bf t','FontSize',16)
        ylabel('\bf x(t)','FontSize',16)
    else
        error('NO CONVERGENCE FOR DESCENT METHOD')
    end
    grad = grad' ;
    cost = cost' ;
    save grad.txt grad -ascii % save vector grad into file grad.txt
    save cost.txt cost -ascii % save vector cost into file cost.txt
    disp('END OF JOB')
```

The program is now complete. We can now improve it, making a slightly faster version. We notice that in the loop for the Armijo method (*for count* ...) we compute the pair $[u, x1]$ and the corresponding cost functional value cv. If the convergence is obtained then $flag2$ gets the value 1 and next we have the sequence

```
if flag2 == 1
    cvnew = cv ;
    ...
    unew = u ;
    ...
end
```

If the outer loop of the gradient algorithm (*for iter* ...) continues, because no convergence criterion was satisfied yet, then we prepare the next iteration by

```
uold = unew ;
cvold = cvnew ;
```

The new iteration begins by solving (3.13) for *uold* (Step S1). But it is clear now that we have already computed the corresponding state for u and it is exactly $x1$. Hence it is no longer necessary to solve (3.13) once again. We therefore modify the program as follows.

```
for iter = 1:maxit
    iter
    if iter == 1
        % S1 : solve (3.13) by Runge–Kutta method
        ...
        jj = 1 ;
    end
    ...
    uold = unew ;
    cvold = cvnew ;
    x = x1 ;
end % for iter
```

We have made two modifications to *CONT0.m*, namely

- The domain of the statement *if iter == 1* was extended upward to include also the sequence for solving the state equation (3.13).

- We have included the statement

    $x = x1$;

    as the last statement in the domain of *for iter* ....

Let us also point out that the above program has four stopping criteria (SC1 to SC4). For a numerical test we can first run the program with all of them. If convergence is obtained we can run the program again with stopping criteria removed or (re)introduced to understand the behavior of the algorithm better.

A first numerical test was made taking $L = 1$, $\alpha = 2$, $\beta = 0.7$, and $a = 3$. As numerical parameters we have considered $h = 0.001$, the initial value for ρ (*roinit*) equal to 1, the maximum number of iterations for the Armijo method (*maxro*) to be 20, and the coefficient b to change ρ (*bro*) equal to 0.55. The precision for the stopping criteria was $\varepsilon = \varepsilon_1 = 0.001$. Taking $x_0 = 0$ in (3.13) the optimal pair is (u^*, x^*) with $u^*(t) = 0$ and $x^*(t) = 0$ for $t \in [0, L]$. Moreover the optimal value is 0. As already mentioned in Section 3.1 the algorithm is convergent to the optimal control (pair) because the cost functional is strictly convex. Even in such a case the behavior of the program can be quite interesting. Taking $u0 = 10$ (as the starting value for the control), at the first iteration $[u^{(0)}, x^{(0)}]$ we are far away from the optimal pair. The program converged numerically to the mathematical optimal pair. The convergence was obtained as follows.

- SC2 (cost criterion), after 15 iterations
- SC3 (control criterion), after 31 iterations
- SC1 (gradient criterion), after 34 iterations

The value of the cost functional after 34 iterations was 2.984×10^{-10}.

Another numerical test was made with the same values as above changing only the starting value for the control to $u0 = 25$. The convergence was obtained as follows.

- SC2 (cost criterion), after 17 iterations
- SC3 (control criterion), after 34 iterations
- SC1 (gradient criterion), after 37 iterations

The value of the cost functional after 37 iterations was 1.92×10^{-10}. What can we notice after these tests?

- The program needs a large number of iterations to approach the optimal pair.
- The convergence criteria are satisfied after different numbers of iterations.
- For $u0 = 25$ the program needs more iterations for convergence.
- Take a look at Figure 3.1. Here we have the optimal control obtained for $u0 = 25$. It looks "interesting" but notice that the GUI (Graphical User Interface) has multiplied all values by 10^5 for graphical purposes. Otherwise all values are "close" to zero.

To take a look at the "speed" of convergence to zero of the algorithm, we give the cost functional values for some iterations corresponding to the initial values $u_0 = 10$ and $x_0 = 0$.

iteration	cost
1	197.1929
2	82.1698
3	34.2942
4	14.3325
5	6.0014
6	2.5172
7	1.0583
8	0.4458
9	0.1883
10	0.0798
11	0.0339
12	0.0144
13	0.0062
14	0.0027
15	0.0011
16	0.0005

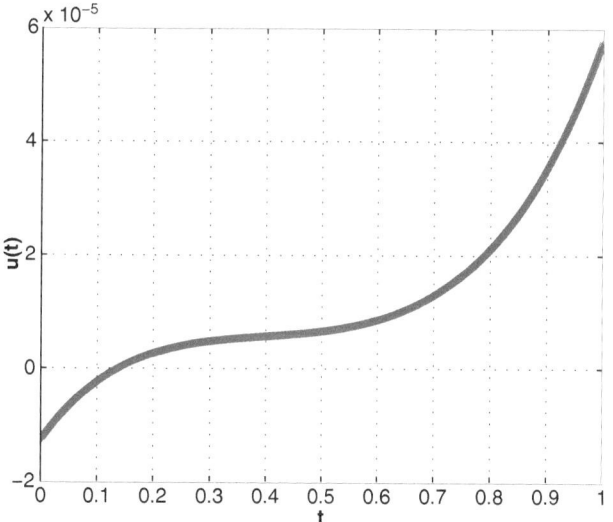

Fig. 3.1. The optimal control for $x_0 := 0$

Another experiment was made changing only $x(0) = 2$ and $u0 = 5$. The convergence was obtained by SC4 after 21 iterations. The corresponding control $u(t)$ is given in Figure 3.2.

In fact it is possible to calculate the optimal control by using the maximum principle. Let (u^*, x^*) be the optimal pair, and the corresponding adjoint state p^*. Hence x^* and p^* are solutions to

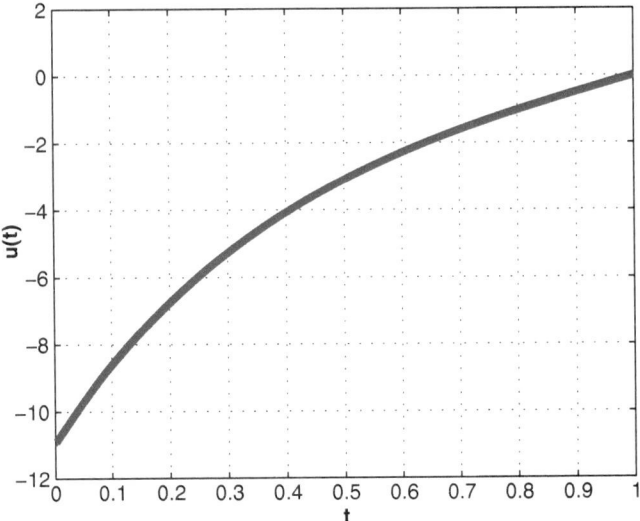

Fig. 3.2. The optimal control for $x_0 := 2$

$$\begin{cases} (x^n)'(t) = \alpha x^*(t) + \beta u^*(t), \quad t \in (0, L) \\ x^*(0) = x_0, \end{cases} \tag{3.19}$$

$$\begin{cases} (p^*)'(t) = -\alpha p^*(t) + a x^*(t), \quad t \in (0, L) \\ p^*(L) = 0. \end{cases}$$

In addition we have that

$$\beta p^*(t) = u^*(t) \quad \text{a.e. } t \in (0, L).$$

It means that u^* may be identified (in $L^2(0, L)$) with an absolutely continuous function βp^*. So, we put βp^* instead of u^* and obtain

$$\begin{cases} (u^*)'(t) = -\alpha u^*(t) + a\beta x^*(t), \quad t \in (0, L) \\ u^*(L) = 0, \end{cases} \tag{3.20}$$

which implies that

$$x^* = \frac{1}{a\beta}((u^*)' + \alpha u^*).$$

We can eliminate x^* and obtain (by (3.19) and (3.20)) that u^* is the solution to

$$\begin{cases} (u^*)''(t) - (\alpha^2 + a\beta^2)u^*(t) = 0, \, t \in (0, L) \\ (u^*)'(0) + \alpha u^*(0) = a\beta x_0 \\ u^*(L) = 0. \end{cases}$$

The solution to this problem is given by

$$u^*(t) = c_1 e^{rt} + c_2 e^{-rt} \quad \text{a.e. } t \in (0, L),\tag{3.21}$$

where $r = \sqrt{\alpha^2 + a\beta^2}$ and

$$\begin{cases} c_1 = \dfrac{a\beta x_0}{(r+\alpha) + (r-\alpha)e^{2rL}}, \\[4mm] c_2 = -\dfrac{a\beta x_0 e^{2rL}}{(r+\alpha) + (r-\alpha)e^{2rL}}. \end{cases}$$

We take the data from the last numerical experiment. If we plot the graph of optimal control given by (3.21) we see that it overlaps to the graph of the approximated optimal control.

3.3 Stock management

We consider here a stock management problem and search for an optimal policy. Let us consider a company that has a delivery contract for a product for the time interval $[0, T]$ $(T > 0)$. According to the contract, the delivery plan is given by a known function $g : [0, T] \rightarrow \mathbb{R}^+$, piecewise continuous. Therefore $g(t)$ is the amount of product to be delivered at moment t. Denote by $u(t)$ the production amount of the company at moment t and by $x(t)$ the stock. The evolution of the stock is described by

$$\begin{cases} x'(t) = u(t) - g(t), \quad t \in (0, T) \\ x(0) = x_0 \in \mathbb{R}. \end{cases}\tag{3.22}$$

The stock is given by

$$x(t) = x_0 + \int_0^t [u(s) - g(s)]ds.$$

- If $x(t) \geq 0$ there is no delivery delay at moment t.
- If $x(t) > 0$ the company has an active stock.
- If $x(t) < 0$ there is a delivery delay at moment t. The company should provide a quantity $|x(t)|$ of product.

We introduce a function $\psi(x)$ which represents the cost of stock conservation or the penalties for the delivery delay. More exactly, $\psi(x)$ = stock conservation price for $x \geq 0$ and $\psi(x)$ = penalties for delivery delays for $x < 0$. We consider that

$$\psi(x) = \begin{cases} c_1 x^2, & x \geq 0 \\ c_2 x^2, & x < 0, \end{cases}\tag{3.23}$$

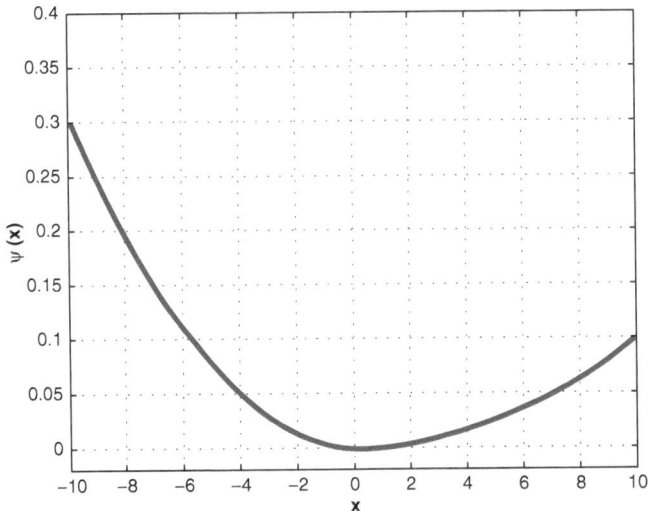

Fig. 3.3. Stock conservation cost/delay penalties

where $c_1 > 0$ and $c_2 > 0$. For example, if we take $c_1 = 0.001$ and $c_2 = 0.003$ the graph of $\psi(x)$ for $x \in [-10, 10]$ is given in Figure 3.3. The values of the parameters mean that the delay is more expensive than an equivalent stock. If $a > 0$ is the unit production price, then the total expenses of the company are given by

$$\Phi(u) = \int_0^T \left[a u(t) + \psi(x^u(t)) \right] dt,$$

where x^u is the solution to (3.22). Therefore a natural optimal control problem is the following.

$$\text{Minimize } \Phi(u), \qquad\qquad\qquad \textbf{(PS)}$$

subject to $u \in K = \{ w \in L^2(0, T); \ u_1 \leq u(t) \leq u_2 \text{ a.e. } t \in (0, T) \}$, where $u_1, u_2 \in \mathbb{R}$, $0 \leq u_1 < u_2$. The control constraints show that the production is bounded. Moreover, for obvious reasons we suppose that

$$u_1 \leq g(t) \leq u_2, \quad t \in [0, T].$$

Problem (PS) is equivalent to the following one

$$\text{Maximize} \{ -\Phi(u) \}, \qquad\qquad\qquad \textbf{(PS')}$$

subject to $u \in K$.

For any arbitrary $u \in U = L^2(0, T)$, consider p^u (the adjoint state) the solution to the following problem,

$$\begin{cases} p'(t) = -\psi'(x^u(t)), & t \in (0,T) \\ p(T) = 0. \end{cases} \tag{3.24}$$

Let us evaluate the gradient of the cost functional.

For any $u, V \in U$ and $\varepsilon \in \mathbb{R}^*$ we have

$$\frac{1}{\varepsilon}[\Phi(u+\varepsilon v) - \Phi(u)] = \int_0^T \frac{1}{\varepsilon}\left[\psi(x^{u+\varepsilon v}(t)) - \psi(x^u(t))\right]dt + a\int_0^T v(t)dt. \tag{3.25}$$

Consider z the solution of

$$\begin{cases} z'(t) = v(t), & t \in (0,T) \\ z(0) = 0. \end{cases} \tag{3.26}$$

It follows as in Sections 2.2 and 2.3 that

$$x^{u+\varepsilon v} \longrightarrow x^u \quad \text{in } C([0,T]),$$

and that

$$\frac{1}{\varepsilon}(x^{u+\varepsilon v} - x^u) \longrightarrow z \quad \text{in } C([0,T]),$$

as $\varepsilon \to 0$. We pass to the limit in (3.25) and we obtain

$$(v, \Phi_u(u))_{L^2(0,T)} = a\int_0^T v(t)dt + \int_0^T \psi'(x^u(t))z(t)dt. \tag{3.27}$$

We multiply the differential equation in (3.24) by z and we integrate on $[0,T]$. We obtain

$$\int_0^T (p^u)'(t)z(t)dt = -\int_0^T \psi'(x^u(t))z(t)dt.$$

We integrate by parts and use (3.24) and (3.26). We get

$$\int_0^T \psi'(x^u(t))z(t)dt = \int_0^T p^u(t)v(t)dt.$$

By (3.27) we get

$$(v, \Phi_u(u))_{L^2(0,T)} = \int_0^T v(t)[p^u(t) + a]dt.$$

Because $v \in L^2(0,T)$ is arbitrary, we may conclude that

$$\Phi_u(u) = p^u + a. \tag{3.28}$$

We are able now to write a descent algorithm for (PS) using $-\Phi_u(u)$ as the descent direction for the current control u. We have to use a projected gradient method (Uzawa's algorithm) because of the control constraints.

A projected gradient algorithm for problem (PS)

S0: Choose $u^{(0)} \in K$.
    Set j:=0.
S1: Compute $x^{(j)}$ the solution to Problem (3.22) with the input $u^{(j)}$:

$$\begin{cases} x'(t) = u^{(j)}(t) - g(t), & t \in (0, T) \\ x(0) = x_0. \end{cases}$$

S2: Compute $p^{(j)}$, the solution to Problem (3.24) with the input $x^{(j)}$:

$$\begin{cases} p'(t) = -\psi'(x^{(j)}(t)), & t \in (0, T) \\ p(T) = 0. \end{cases}$$

S3: Compute the gradient direction $w^{(j)}$ using formula (3.28):

$$w^{(j)} := \Phi_u(u^{(j)}) = p^{(j)} + a.$$

S4: (The stopping criterion)
    If $\|\Phi_u(u^{(j)})\| < \varepsilon$
        then STOP ($u^{(j)}$ is the approximating control)
        else go to S5;
S5: Compute the steplength ρ_j such that

$$\Phi(P_K(u^{(j)} - \rho_j w^{(j)})) = \min_{\rho \geq 0}\{\Phi(P_K(u^{(j)} - \rho w^{(j)}))\}.$$

S6: Compute the new control

$$u^{(j+1)} := P_K(u^{(j)} - \rho_j w^{(j)});$$

$j := j + 1$; go to S1.

The projection operator $P_K : L^2(0, T) \to K$ may be defined pointwise with respect to t; that is,

$$P_K(u)(t) = Proj(u(t)) \text{ a.e. } t \in (0, T),$$

where

$$Proj(w) = \begin{cases} w & \text{if } u_1 \leq w \leq u_2 \\ u_1 & \text{if } w < u_1 \\ u_2 & \text{if } w > u_2. \end{cases}$$

As asserted before in Section 3.2 such an algorithm can be provided with different stopping criteria (SC1 to SC4). There we wrote the conceptual algorithm with SC2. Here we use SC1 just because it takes a different position inside the algorithm. For the program SC2 and SC3 were used. The design of the program is quite similar to the one in Section 3.2. This is why we give there only the sequences that are different. We start by PART1. We declare as global variables: ind, L2, c1, c2, u1, u2, g1, g2, g3, g4, h, uold, u, and x.

Here L is the final time (T) and

L2 = L/2 ; % the midinterval for function g

We then read the corresponding data and make the initializations. For our numerical tests we have chosen

$$g(t) = \begin{cases} g_1 t + g_2, & t \in [0, T/2] \\ g_3 - g_4 t, & t \in (T/2, T]. \end{cases}$$

By taking

$$g_1 = g_4 = \frac{2}{T}(u_2 - u_1), \quad g_2 = u_1, \quad g_3 = 2u_2 - u_1,$$

the function $g(t)$ satisfies the constraints $u_1 \leq g(t) \leq u_2$. Moreover, it is continuous on $[0, T]$ and $g(0) = g(T) = u_1$, $g(T/2) = u_2$. This is a normal delivery plan for a company. In our program the variable L stands for T and this is why we have introduced the statement $L2 = L/2$.

We continue the program also including the "shortcut" from Section 3.2.

```
% ============
%      PART 2
% ============
% gradient loop starts
for iter = 1:maxit
      iter
      if iter == 1
            % S1 : solve (3.22) by RK4 method
            x(1) = x0 ;
            for i = 1:n − 1
                  ind = i ;
                  x(i + 1) = RK41(x(i),t(i)) ;
            end
            disp('State Problem solved') ;
            for i = 1:n
                  z(i) = f(x(i)) ; % function f stands for ψ
            end
            Q = a*uold + z ;
            temp = trapz(t,Q) ;
            cvold = temp
            cost(1) = cvold ;
            jj = 1 ;
      end % end if
```

The file function *RK41.m* remains unchanged, but the inside function *F1.m* has changed because the state equation is now different from the one in Section 3.2. We therefore have

```
function yout = F1(t,x)
global ind
global uold
yout = uold(ind) − g(t) ;
```

According to the form of function g, the function file *g.m* is

```
function y = g(t)
global L2
global g1 g2 g3 g4
if t <= L2
    y = g1*t + g2 ;
else
    y = g3 − g4*t ;
end
```

The function file *f.m* that computes the function ψ is

```
function y = f(x)
global c1 c2
temp = x*x ;
if x >= 0
    y = c1*temp ;
else
    y = c2*temp ;
end
```

PART 3 of the program remains unchanged with the exception of the gradient formula which is different. We therefore have

```
% S3 : compute the gradient
w = p + a ;
disp('GRADIENT COMPUTED') ;
```

The adjoint equation is different from the one in Section 3.2 therefore we should also change the inside function file *F3.m* from *RK43.m*. We have

```
function yout = F3(t,p)
global ind
global x
yout = −fder(x(ind)) ;
```

We have to write the file *fder.m* corresponding to the derivative of function ψ given by formula (3.23). *PART 4* of the program (test of SC1) remains unchanged. A modification appears in *PART 5* inasmuch as we deal with the projected gradient method. The computation of the intermediate control u becomes

```
for count = 1:maxro
    count
    for i = 1:n
        temp = uold(i) − robar*w(i) ;
        u(i) = Proj(temp) ;
    end
```

The function file *Proj.m*, which applies the formula of the projection operator P_K, follows.

```
function y = Proj(u)
global u1 u2
y = u ;
if u < u1
    y = u1 ;
end
if u > u2
    y = u2 ;
end
```

The function file *RK42.m* remains unchanged (the Runge–Kutta formula is the same) but the inside function *F2.m* becomes

```
function yout = F2(t,x)
global ind
global u
yout = u(ind) − g(t) ;
```

Another modification in *PART 5* corresponds to the computation of state vector $x1$ for the control vector u and of the corresponding cost value.

```
for j = 1:n
    z1(j) = f(x1(j)) ;
end
Q1 = a*u + z1 ;
temp = trapz(t,Q1) ;
cv = temp ;
```

We now pass to a first numerical example.

Example 1. We take $L = 12$ (months), $a = 0.1$, $x(0) = 5$, $u_1 = 10$, $u_2 = 16$, and $u(0) = 14$. The coefficients of function $\psi(x)$ are $c_1 = 0.001$ and $c_2 = 0.003$, which means that delivery delay is more expensive than positive stock. The coefficients of function $g(t)$ are $g_1 = g_4 = 1$, $g_2 = 10$, and $g_3 = 22$. The numerical parameters for the gradient algorithm are $h = 0.001$ and $\varepsilon = 0.001$. For the Armijo method we take *roinit* $= 1$, $b = 0.6$, $\varepsilon_1 = 0.001$, and

$maxro = 20$. The approximating optimal control is represented in Figure 3.4. The corresponding value of the cost functional is 14.7832.

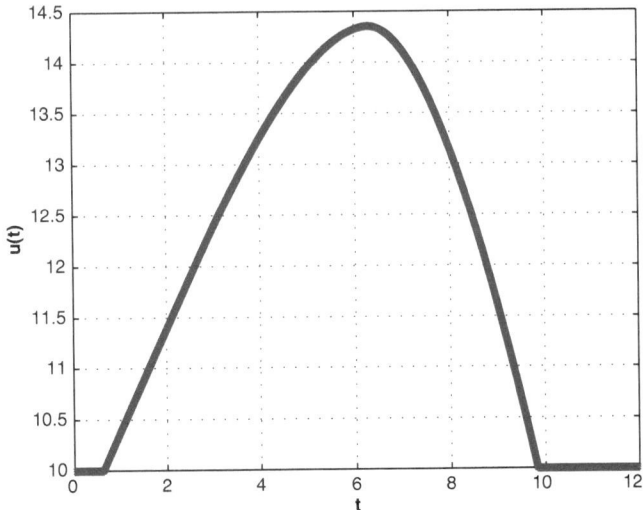

Fig. 3.4. The control for $x(0) = 5$, $c_1 = 1e-3$, $c_2 = 3e-3$

Example 2. We take all values as in Example 1 modifying only the coefficients of the pay function $\psi(x)$ to be $c_1 = 0.004$ and $c_2 = 0.001$. This means that a positive stock is more expensive than the delivery delay. The control is given in Figure 3.5. The corresponding value of the cost functional is 14.4741.

It is quite obvious that the preference for delivery delay (given by the coefficients c_1 and c_2) is exploited in Example 2 in comparison to Example 1.

Example 3. We take once again all values as in Example 1 except the initial stock value which is $x(0) = -5$. The result is given in Figure 3.6. The corresponding value of the cost functional is 15.762.

Let us notice that the negative stock at time $t = 0$ ($x(0) = -5$) implies higher values for the production $u(t)$ (see the values on the Oy-axis), at least in the first part of the time interval, in comparison to Example 1 where $x(0) = 5$.

3.4 An optimal harvesting problem

We consider a simple harvesting problem ([BG80, Sections 8.2.3 and 12.4]). More exactly, let $x(t)$, with $t \in [0, L]$, be a renewable population (e.g., fish

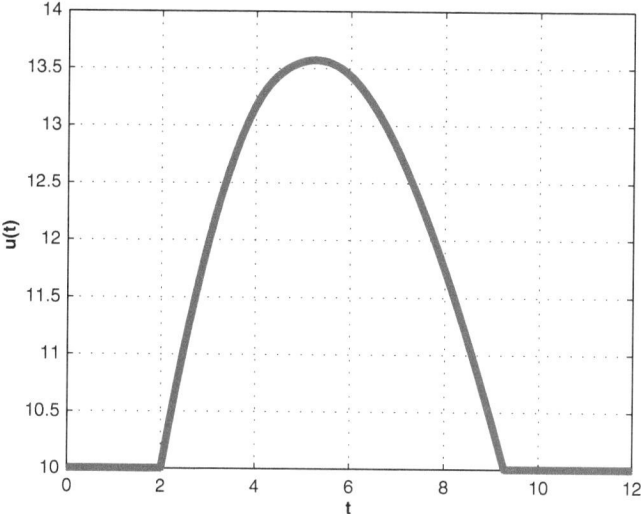

Fig. 3.5. The control for $x(0) = 5$, $c_1 = 4e - 3$, $c_2 = 1e - 3$

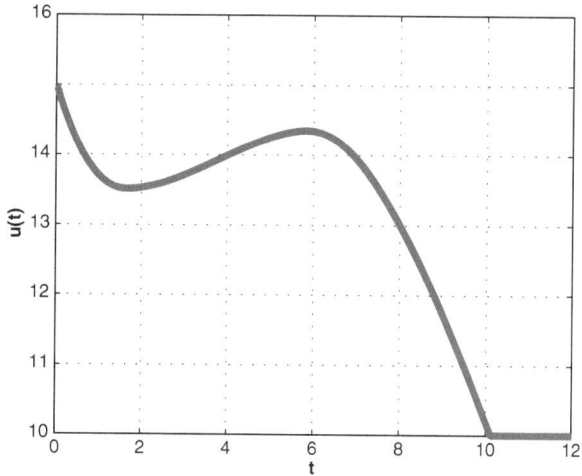

Fig. 3.6. The control for $x(0) = -5$, $c_1 = 1e - 3$, $c_2 = 3e - 3$

population, grazing land, forest). We are interested in its optimal management. We consider the equation

$$x'(t) = F(x(t)) - h(t), \quad t \in (0, L), \tag{3.29}$$

where F is the growth law and $h(t)$, is the harvesting rate. We take the Verhulst version of the growth function F; that is,

$$F(x) = rx \left(1 - \frac{x}{k} \right). \tag{3.30}$$

Here $r > 0$ is the intrinsic growth rate and $k > 0$ is the carrying capacity of the environment (see Section 1.5 for a similar model). Moreover,

$$h(t) = u(t)x(t),$$

where u, the fishing effort, is proportional to the population level. We want to maximize the economic rent which is defined as the difference between revenue and cost. We have $revenue = ah$, where $a > 0$ is the unit price, and $cost = cu$, where $c > 0$ is a constant: the cost is proportional to the fishing effort. Therefore the control problem is

$$\text{Maximize } \Phi(u) = \int_0^L e^{-\delta t}(aux^u - cu)dt \qquad \textbf{(PH)}$$

subject to

$$u \in K = \{w \in L^2(0, L); \ 0 \le w(t) \le \bar{u} \text{ a.e. } t \in (0, L)\}, \quad \bar{u} > 0,$$

where x^u is the solution to the state problem:

$$\begin{cases} x'(t) = F(x(t)) - u(t)x(t), & t \in (0, L) \\ x(0) = x_0 > 0. \end{cases} \qquad (3.31)$$

The control restriction means that the fishing effort is bounded. Moreover, $\delta > 0$ is the discount factor. The solution to (3.31) is of course positive. Let (u^*, x^*) be the optimal pair (the existence and the uniqueness follow as in Sections 2.2 and 2.3). Consider an arbitrary but fixed $u \in K$ and $v \in V = \{w \in L^2(0, L); \ u+\varepsilon w \in K \text{ for any } \varepsilon > 0 \text{ small enough}\}$. We start to compute the gradient $\Phi_u(u)$. Let $z = z^u$, the solution to

$$\begin{cases} z'(t) = F'(x^u(t))z(t) - u(t)z(t) - v(t)x^u(t), & t \in (0, L) \\ z(0) = 0. \end{cases} \qquad (3.32)$$

We calculate $\varepsilon^{-1}[\Phi(u + \varepsilon v) - \Phi(u)]$, we pass to the limit for $\varepsilon \to 0+$, and we obtain

$$(v, \Phi_u(u)) = \int_0^L e^{-\delta t}(au(t)z(t) + ax^u(t)v(t) - cv(t))dt. \qquad (3.33)$$

We introduce the adjoint state $p = p^u$ which satisfies the adjoint equation

$$\begin{cases} p'(t) = -e^{-\delta t}au(t) - [F'(x(t)) - u(t)]p(t), & t \in (0, L) \\ p(L) = 0. \end{cases} \qquad (3.34)$$

We multiply the differential equations in (3.32) by p^u and we integrate by parts on $[0, L]$. Next we multiply the differential equations in (3.34) by z and we integrate on $[0, L]$, and comparing the two formulae we obtain that

$$\int_0^L e^{-\delta t}auzdt = -\int_0^L x^u p^u vdt.$$

Replacing into (3.33) we find that

$$\Phi_u(u) = e^{-\delta t}(ax^u - c) - x^u p^u. \tag{3.35}$$

The optimal control u^* satisfies in addition

$$\Phi_u(u^*) = e^{-\delta t}(ax^{u^*} - c) - x^{u^*} p^{u^*} \in N_K(u^*). \tag{3.36}$$

Therefore (we denote $x^* = x^{u^*}$ and $p^* = p^{u^*}$):

- If $e^{-\delta t}(ax^* - c) - x^* p^* > 0$, then $u^*(t) = \bar{u}$.
- If $e^{-\delta t}(ax^* - c) - x^* p^* < 0$, then $u^*(t) = 0$.

We analyze what happens with $u^*(t)$ when t belongs to the set $B = \{t \; ; \; e^{-\delta t}(ax^* - c) - x^* p^* = 0 \quad \text{a.e.}\}$. From the definition of B we get p^* and by derivation, and combining with (3.34) we obtain the equation satisfied by x^*:

$$F'(x) + \frac{cF(x)}{x(ax - c)} = \delta. \tag{3.37}$$

Replacing (3.30) into (3.37) yields a quadratic equation for x^*, namely

$$\alpha x^2 + \beta x + \gamma = 0 \; ,$$

where

$$\alpha = \frac{2ar}{k} > 0, \quad \beta = a(\delta - r) - \frac{cr}{k}, \quad \gamma = -c\delta < 0.$$

A simple analysis shows that the above equation has two real roots, one of them being positive and the other negative. We denote by $\tilde{x}$ the positive one. In conclusion $x^*(t) = \tilde{x}$ on B and from (3.31) we obtain the corresponding control $\tilde{u} = F(\tilde{x})/\tilde{x}$. Therefore the optimal control takes only the values $\{0, \tilde{u}, \bar{u}\}$. Moreover, natural conditions on the coefficients ensure that $0 < \tilde{u} < \bar{u}$. A simple calculation shows that $0 < \tilde{x} < k$. Hence $F(\tilde{x}) > 0$ and this implies $\tilde{u} > 0$. On the other hand $\tilde{u} < \bar{u}$ means $F(\tilde{x})/\tilde{x} < \bar{u}$ and we readily get as sufficient condition $\bar{u} > r$.

In [BG80] the following control is recommended as an "optimal" one.

$$u(t) = \begin{cases} \bar{u}, & x^*(t) > \tilde{x} \\ \tilde{u}, & x^*(t) = \tilde{x} \\ 0, & x^*(t) < \tilde{x}. \end{cases} \tag{3.38}$$

The control defined by formula (3.38) is not optimal, in fact, as can be emphasized by the numerical tests. This control has the property to "bring" the population $x(t)$ to the level $\tilde{x}$. We write a program for it to show this interesting feature and to calculate the corresponding value of the cost functional used later.

```
% file harv1.m
% a harvesting problem
```

```
clear
global r k
global util ubar xtil eps1
% Introduce below input statements for
% L, y0 (for x(0)), r, k, a, c, d (for δ), ubar (for ū), eps1 (for ε₁), h.
% compute xtil=x̃ and util=ũ
alf = 2*a*r/k ;    % this is α
b = a*(d − r) − c*r/k ;    % this is β
cd = −c*d ;    % this is γ
w = [alf,b,cd]    % this is the polynomial
x = roots(w)
if x(1) > 0
    xtil = x(1) ;
else
    xtil = x(2) ;
end
xtil
util = F(xtil) / xtil
% end of sequence for util and xtil
tspan = 0:h:L ;
[t y] = ode45('rhs',tspan,y0) ;
plot(t,y,'r*') ; grid
xlabel('\bf t','FontSize',16)
ylabel('\bf x(t)','FontSize',16)
n = length(t) ;
u = zeros(n,1) ;
w = zeros(n,1) ;
% compute the control ũ
for i = 1:n
    if y(i) >= xtil + eps1
        u(i) = ubar ;
    end
    if abs(y(i)-xtil) < eps1
        u(i) = util ;
    end
end
figure(2)
plot(t,u,'*') ; grid
xlabel('\bf t','FontSize',16)
ylabel('\bf u(t)','FontSize',16)
for i = 1:n
    w(i) = exp(−d*t(i))*(a*y(i) − c)*u(i) ;
end
cost = trapz(t,w)
```

We now give the function file *rhs.m*.

```
function out1 = rhs(t,y)
global util ubar xtil eps1
temp = 0 ;
if abs(y − xtil) < eps1
    temp = util ;
end
if y >= xtil + eps1
    temp = ubar ;
end
out1 = F(y) − temp*y ;
```

Here and in the script file we have to handle a current problem in scientific calculus. The problem is to apply formula (3.38). To translate the mathematical test $x(t) = \tilde{x}$ by $y == xtil$ in the program is meaningless because of computer arithmetic and roundoff errors. A correct numerical test is $|x(t) - \tilde{x}| < \varepsilon_1$, where $\varepsilon_1 > 0$ is a prescribed precision (variable *eps1* in the program). This means that formula (3.38) is replaced by

$$u(t) = \begin{cases} \bar{u}, & x(t) \geq \tilde{x} + \varepsilon_1 \\ \tilde{u}, & x(t) \in (\tilde{x} - \varepsilon_1, \tilde{x} + \varepsilon_1) \\ 0, & x(t) \leq \tilde{x} - \varepsilon_1. \end{cases} \tag{3.39}$$

The function file *F.m* contains the corresponding function from formula (3.30). It looks like

```
function y = F(x)
global r k
y = r*x*(1−x/k) ;
```

We give a numerical test here. We take $L = 1$, $r = 0.3$, $k = 5$, $a = 1$, $c = 1$, $\delta = 0.5$, and $\varepsilon = 0.001$. We obtain $\tilde{x} = 1.5396$ and $\tilde{u} = F(\tilde{x})/\tilde{x} = 0.2076$. We therefore take $\bar{u} = 0.5$. The states (populations) corresponding to $x(0) = 1.4$ and to $x(0) = 1.6$ are given in Figure 3.7. Notice that both trajectories are "convergent" to the value $\tilde{x}$.

We should also assert that the parameter ε_1 (*eps1*) is "slippery". We mean that its value should be chosen very carefully. For instance, for tiny values the computation is no longer stable. Of course ε_1 also should not be too large. Here $\varepsilon_1 = 0.001$ is a suitable value. Later we discuss this problem for a numerical example.

We now pass to the numerical approximation of the optimal value of the cost functional. Because our control problem is a maximizing one we use the gradient direction $\Phi_u(u^{(j)})$ for the current control $u^{(j)}$. The algorithm follows. As in the previous section we write only the stopping criterion SC1. We also show later that the behavior of the gradient method is a little more complicated for this example.

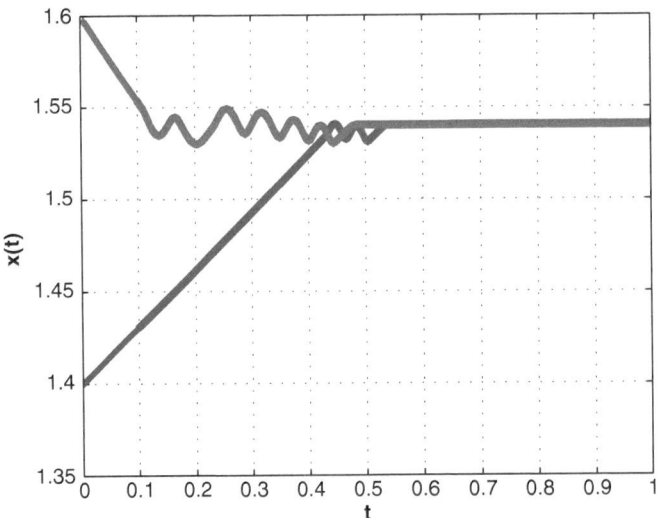

Fig. 3.7. Population evolution for harv1.m

A projected gradient algorithm for problem (PH)

S0: Choose $u^{(0)} \in K$.
  Set j:=0.
S1: Compute $x^{(j)}$ the solution to Problem (3.31) with the input $u^{(j)}$:

$$\begin{cases} x'(t) = F(x(t)) - u^{(j)}(t)x(t), \quad t \in (0, L) \\ x(0) = x_0. \end{cases}$$

S2: Compute $p^{(j)}$, the solution to Problem (3.34) with the input $x^{(j)}$:

$$\begin{cases} p'(t) = -e^{-\delta t}au^{(j)}(t) - \big(F'(x^{(j)}(t)) - u^{(j)}(t)\big)p(t), \quad t \in (0, L) \\ p(L) = 0. \end{cases}$$

S3: Compute the gradient direction $w^{(j)}$ using formula (3.36):

$$w^{(j)} := \Phi_u(u^{(j)}) = e^{-\delta t}(ax^{(j)} - c) - x^{(j)}p^{(j)}.$$

S4: (The stopping criterion)
  If $\|w^{(j)}\| < \varepsilon$
    then STOP ($u^{(j)}$ is the approximating control)
    else go to S5;
S5: Compute the steplength ρ_j such that

$$\Phi(P_K(u^{(j)} + \rho_j w^{(j)})) = \max_{\rho \geq 0} \{\Phi(P_K(u^{(j)} + \rho w^{(j)}))\}.$$

S6: Compute the new control

$$u^{(j+1)} := P_K(u^{(j)} + \rho_j w^{(j)});$$

$j := j + 1$; go to S1.

The projection operator $Proj$, may be defined pointwise with respect to t as discussed in previous sections; that is,

$$(Proj(u))(t) = Proj(u(t)) \text{ a.e. } t \in [0, L].$$

Moreover the pointwise projection operator should be defined $Proj : \mathbb{R} \to \{0, \tilde{u}, \bar{u}\}$ because, as demonstrated earlier, optimal control takes only the values 0, $\tilde{u}$, and $\bar{u}$. Therefore we use the minimum distance to one of the three points. To simplify the explanation we give the corresponding file function $Proj2.m$ directly (notice that in this function file $util$ stands for $\tilde{u}$ and $ubar$ for $\bar{u}$).

```
function y = Proj2(u)
global ubar util
if u <= 0
    y = 0 ;
    return
end
if u >= ubar
    y = ubar ;
    return
end
z(1) = u ;
z(2) = abs(u − util) ;
z(3) = abs(u − ubar) ;
[zm,j] = min(z) ;
if j == 1
    y = 0 ;
    return
end
if j == 2
    y = util ;
    return
end
y = ubar ;
```

As we explain later we have also used the classical projection from $\mathbb{R}$ to $[0, \bar{u}]$ in order to improve the behavior of the algorithm (program). We also give the corresponding function file $Proj.m$.

```
function y = Proj(u)
global ubar
y = u ;
if u < 0
    y = 0 ;
end
if u > ubar
    y = ubar ;
end
```

The numerical tests are tedious for this problem. We have used more than a program:

- *harv1.m* to compute the cost value corresponding to the control defined by formula (3.39).
- *harv2.m* to compute the optimal pair (with *Proj.m* and stopping criteria SC1, SC2, and SC3).
- *harv35.b* to compute the optimal pair (with *Proj2.m* and stopping criteria SC1 and SC4).
- *harv35a.m* to compute the optimal pair using as starting control the optimal one furnished by *harv2.m* *(with Proj2.m* and stopping criteria SC1 and SC4).

Except for *harv1.m*, the other three programs are quite similar, the differences being mentioned above. We now give the differences between *harv2.m* and the tutorial program *CONT0.m* from Section 3.2. Because we take $\delta = 0$ for the numerical tests, the program contains the formulae from the projected gradient algorithm above written for $\delta = 0$.
We start with by PART1 which contains statements for global variables (r, k, a, ubar, ind, h, uold, u, x), input statements for data, and also the sequence

```
t = 0:h:L ;
t = t' ;
n = length(t) ;
```

For *harv35b.m* and *harv35a.m*, which use *Proj2.m*, we introduce the sequence from *harv1.m* to compute *xtil* and *util*.

```
% compute xtil=x̃ and util=ũ

.....
% end of sequence for util and xtil
```

The only thing changed in *PART 2* is the computation of the cost functional value which reads

```
disp('SE solved') ;
for i = 1:n
    q(i) = uold(i)*(a*x(i) − c) ;
end
```

```
temp = trapz(t,q) ;
cvold = temp
```

The function file *F1.m* called by *RK41.m* is

```
function yout = F1(t,x)
global ind
global uold
yout = F(x) − uold(ind)*x ;
```

The function file *F.m* containing the function from formula (3.30) was already given above. We now pass to *PART 3*. To solve the adjoint system, the function file *F3.m* called by *RK43.m* becomes

```
function yout = F3(t,p)
global a
global ind
global uold
global x
yout = (p − a)*uold(ind) − Fder(x(ind))*p ;
```

The function file *Fder.m* gives the derivative of the function $F(x)$ from formula (3.30). It looks like

```
function y = Fder(x)
global r k
y = r*(1 − 2*x/k) ;
```

Another change in *PART 3* concerns the computation of the gradient.

```
% S3 : compute the gradient
for i = 1:n
    w(i) = (a − p(i))*x(i) − c ;
end
disp('GRADIENT COMPUTED') ;
```

Other changes are made in *PART 5*. We start with the computation of the steplength.

```
for count = 1:maxro
    count
    for i = 1:n
        temp = uold(i) + robar*w(i) ;
        u(i) = Proj(temp) ; % use Proj2 for harv35a,b.m
end
```

The function file *F2.m* called by *RK42.m* is now

```
function yout = F2(t,x)
global ind
global u
yout = F(x) − u(ind)*x ;
```

The new cost value is computed by

```
for i = 1:n
    q1(i) = u(i)*(a*x1(i) − c) ;
end
temp = trapz(t,q1) ;
cv = temp ;
```

Finally the test for the cost value changes because we are dealing with a maximization problem. For clarity we give the whole program sequence until the end of the *count* loop. Compare it with the corresponding sequence from the program *CONT0.m* from Section 3.2 which is made for a minimization problem.

```
        if cv <= cvold    % no increase of cost for maximization
            robar = bro * robar
        else
            flag2 = 1 ;
            break
        end
    end    % for count
```

The program is complete for each version. We pass to numerical experiments.

Example 1. We take $L = 10$, $r = 0.3$, $k = 5$, $a = 1$, $c = 1$, $\delta = 0$. We therefore obtain $\tilde{x} = 3$ and $\tilde{u} = 0.12$. Hence we choose $x(0) = 2$, $\bar{u} = 1.5$, and $u(0) = 1.4$. We run programs as follows.

- *harv1.m* which uses formula (3.39). We have also tested the parameter ε_1 and we have obtained the following cost values for $h = 0.001$:

ε_1	cost
10^{-1}	1.7646
10^{-2}	1.7537
10^{-3}	1.7665
10^{-4}	0.7232

  Notice the poor result for $\varepsilon_1 = 10^{-4}$. This means that ε_1 is too tiny.
- *harv35b.m* (with $b = 0.6$, $h = 0.001$, $\varepsilon_1 = 10^{-3}$, *maxro* = 20) has obtained after 22 iterations (including ρ modifications) the cost value 1.9724 and the control $u(t)$ from Figure 3.8. The program was not able to improve the cost value for the precision $\varepsilon = 10^{-4}$. This is why next we have used *harv2.m* with *Proj.m*.
- *harv2.m* with the classical projection has obtained after 70 iterations the cost value 2.5767 with convergence by SC2. The corresponding control $u(t)$ is given in Figure 3.9.

Notice that the control $u(t)$ uses in many points the values 0, $\tilde{u} = 0.12$, and $\bar{u}$. At the end we save the optimal control into file *u2.txt* by the statement

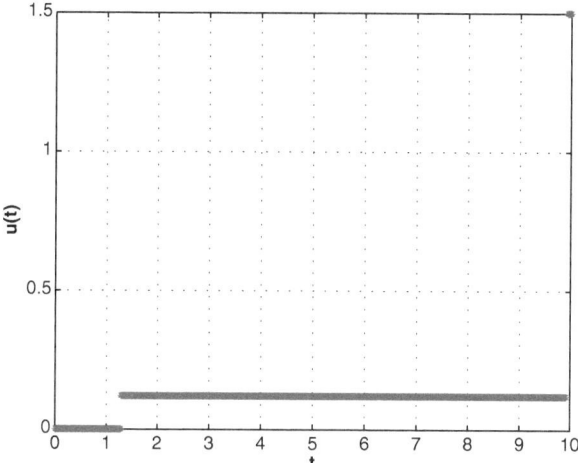

Fig. 3.8. Control by harv35b.m and Proj2.m

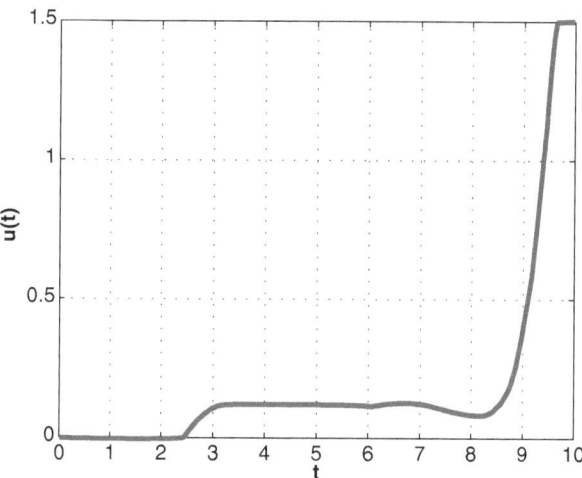

Fig. 3.9. Control by harv2.m and Proj.m

save u2.txt unew -ascii

• *harv35a.m* was started using the control saved by *harv2.m*. This was done by the sequence

```
load u2.txt
for i = 1:n
    uold(i) = u2(i) ;
end
```

After 18 other iterations (including the ones for ρ) we have obtained a better value for the cost functional, namely 2.5791. The corresponding control $u(t)$ is presented in Figure 3.10.

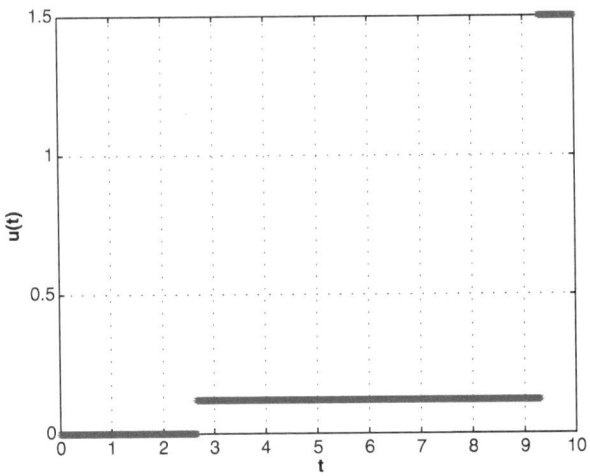

Fig. 3.10. Control by harv35a.m and Proj2.m

Finally we write the numerical results for the cost functional into a centralized table. The increase of the cost values is quite interesting. We have obtained

program   cost
harv1.m   1.7665
harv35b.m 1.9724
harv2.m   2.5767
harv35a.m 2.5791

Example 2. The data are as in Example 1, except for the following changes. We modify the initial values and we take $u(0) = 0.12$ and $x(0) = 3$. With the same program policy, we have obtained the following numerical results.

program   cost
harv1.m   2.4
harv35b.m 2.8989
harv2.m   3.2252
harv35a.m 3.2276

Example 3. The data are as in Example 1, except for the following changes. We now take $L = 5$, $\varepsilon = 10^{-3}$, and the initial values $u(0) = 0.12$ and $x(0) = 2$. With the same program policy, we have obtained the following numerical results.

program	cost
harv1.m	0.5489
harv35b.m	1.3534
harv2.m	1.3726
harv35a.m	1.3789

Bibliographical Notes and Remarks

The first monographs devoted to the (numerical) approximation of optimal control problems are [Pol71] and [Cea71]. Later other books developed many new results such as [Cea78] and [GS81]. We also recommend [LW07].

As asserted in this chapter a gradient method as an iterative procedure has the general form $u^{(k+1)} = u^{(k)} + \rho_k w^{(k)}$, where $u^{(k)}$ is the current control, $w^{(k)}$ is a *search direction*, and ρ_k is the *steplength*. Here we have used only the (opposite) gradient direction for $w^{(k)}$ and the Armijo method to fit ρ_k. They worked well. In the literature above there are other, more sophisticated, choices for $w^{(k)}$ and also for ρ_k. We also cite [AN03, Chapter 2].

For the example in Section 3.2 we refer to [Kno81], for the one in Section 3.3 we cite [CV78]. The example in Section 3.4 is due to [BG80]. The literature dedicated to optimal harvesting problems is quite large. More complex models than the one in Section 3.4 are presented in Chapter 4 of this book. See also the Bibliographical Notes in the next chapter.

Exercises

3.1. We consider the following optimal control problem.

$$\text{Minimize } \frac{1}{2}\int_0^1 (u^2(t) + x^2(t))dt,$$

subject to $u \in L^2(0,1)$, where $x = x^u$ is the solution to

$$\begin{cases} x'(t) = u(t), & t \in (0,1) \\ x(0) = 1. \end{cases}$$

Find the optimality conditions (and the gradient $\Phi_u(u)$) and obtain the optimal control.

Hint. This is a particular case of the problem in Section 3.2. Let p^u be the adjoint state. The adjoint problem is

$$\begin{cases} p'(t) = -x^u(t), & t \in (0,1) \\ p(1) = 0. \end{cases}$$

The gradient is
$$\Phi_u(u) = u + p^u.$$

The optimal control is
$$u^*(t) = -\frac{\sinh(1-t)}{\cosh(1)}.$$

The corresponding optimal state $x^* = x^{u^*}$ is
$$x^*(t) = \frac{\cosh(1-t)}{\cosh(1)},$$

where
$$\sinh(t) = \frac{1}{2}(e^t - e^{-t}),$$

$$\cosh(t) = \frac{1}{2}(e^t + e^{-t}).$$

Write a program to approximate the optimal pair by a gradient method and compare the numerical results with the mathematical ones.

3.2. Add the term $\varphi(x^u(T))$ to the cost functional of Problem (P1′) from Section 3.1. Calculate what is necessary and write an algorithm similar to the gradient-type one in Section 3.1.

3.3. Consider the following optimal control problem,

$$\text{Maximize } \int_0^T u(t)x^u(t)dt - c\int_0^T u(t)^2 dt,$$

subject to $u \in L^2(0,T)$, $0 \le u(t) \le M$ a.e. $t \in (0,T)$, where x^u is the solution of the following logistic model of population dynamics:

$$\begin{cases} x'(t) = rx(t) - kx(t)^2 - u(t)x(t), & t \in (0,T) \\ x(0) = x_0 > 0. \end{cases}$$

Here c, r, k, x_0 are positive constants. Calculate the gradient of the cost functional and write a program using a gradient-type algorithm.

Hint. See Exercise 2.5 in the previous chapter and take into account the control constraints (a projected gradient method is necessary).

3.4. Problem (PS) is a minimization one, whereas (PS′) is a maximization one. Transform the program *stock1.m* from Section 3.3, written for (PS), into a program to solve Problem (PS′) numerically. Make the same numerical tests and notice if you obtain opposite values for the cost functional.

Hint. For the actualization of the cost value (Armijo method) also take a look at program *harv2.m* from Section 3.4 which is written for a maximization problem.

3.5. Consider the optimal control problems from Working Example 1 in Chapter 2. For each of them calculate the gradient of the cost functional and write a program using a gradient-type algorithm.

3.6. Consider the optimal control problems from Working Example 2 in Chapter 2. For each of them calculate the gradient of the cost functional and write a program using a gradient-type algorithm.

4

Optimal harvesting for age-structured population

This chapter is intended to be a bridge towards scientific research on optimal control theory. The problems investigated here are much more complex than those presented in the previous chapters. We focus on optimal harvesting problems for age-structured population dynamics, which are extremely important from a biological as well as from an economical point of view. Even if the degree of complexity of the optimal control problems investigated in this chapter is much higher than before we can see that the steps we have to follow are the same as for the optimal control problems investigated in Chapter 2.

Age is a parameter of great importance in population dynamics. This chapter is devoted to the study of one of the most important age-dependent models describing the dynamics of a single population species. The basic properties of the solutions are investigated. Numerical solutions with MATLAB® are obtained. We also treat an optimal harvesting problem related to the linear age-dependent population dynamics. The optimal harvesting effort is approximated using MATLAB. An optimal control for age-dependent population dynamics with periodic vital rates and logistic term is also investigated.

4.1 The linear age-structured population dynamics

This section is devoted to some basic properties of the solutions to a linear model of population dynamics with an age-structure. The results presented here are extremely important in view of the investigation of optimal harvesting problems.

Throughout this chapter A, $T \in (0, +\infty)$, $Q_T = (0, A) \times (0, T)$ and Dy is the directional derivative of y with respect to direction $(1, 1)$; that is,

$$Dy(a, t) = \lim_{h \to 0} \frac{y(a + h, t + h) - y(a, t)}{h}.$$

S. Aniţa et al. *An Introduction to Optimal Control Problems in Life Sciences and Economics*, Modeling and Simulation in Science, Engineering and Technology, DOI 10.1007/978-0-8176-8098-5_4, © Springer Science+Business Media, LLC 2011

Consider a single biological population species and denote by $y(a,t)$ the number (density) of individuals of age $a \in [0, A]$ at time $t \in [0, T]$. Assume that y is smooth enough.

Let $\mu(a,t)$ be the mortality rate of individuals of age $a \in [0, A]$ at time $t \in [0, T]$. This gives the proportion of dying individuals and depends on age a and time t. The balance law shows that the number of individuals of age a at the moment t that die in the time interval $[t, t + dt]$ is

$$y(a, t) - y(a + dt, t + dt) = \mu(a, t)y(a, t)dt.$$

Dividing now by dt we obtain

$$Dy(a, t) + \mu(a, t)y(a, t) = 0.$$

If a certain infusion of population (inflow) occurs then the population dynamics is described by the equation:

$$Dy(a, t) + \mu(a, t)y(a, t) = f(a, t), \quad (a, t) \in Q_T, \tag{4.1}$$

where A is the maximal age for the population species and $f(a, t)$ is the density of the inflow of population of age a at time t.

Equation (4.1) has been proposed by F. R. Sharpe, A. Lotka, and by A. G. McKendrick (see, e.g., [Ani00], [Ian95], [Thi03], and [Web85]). If y is sufficiently smooth, then

$$Dy(a, t) = \frac{\partial y}{\partial a} + \frac{\partial y}{\partial t};$$

that is why in many papers Sharpe–Lotka–McKendrick's equation appears as

$$\frac{\partial y}{\partial a}(a, t) + \frac{\partial y}{\partial t}(a, t) + \mu(a, t)y(a, t) = f(a, t),$$

which is a first-order partial differential equation.

The birth process is described by the renewal law:

$$y(0, t) = \int_0^A \beta(a, t)y(a, t)da, \quad t \in (0, T), \tag{4.2}$$

where $y(0, t)$ is the number of newborns at time t and $\beta(a, t)$ is the fertility rate and gives the proportion of newborn population at moment t with parents of age a. Moreover, (4.2) is a nonlocal boundary condition.

The initial distribution of the population is given by

$$y(a, 0) = y_0(a), \quad a \in (0, A), \tag{4.3}$$

where y_0 is a known function.

In real situations it has been observed that for any fixed $t \geq 0$, the graphs of $\beta(\cdot, t)$ and $\mu(\cdot, t)$ are as in Figure 4.1 (at the end of this section). The graphs correspond to $\beta(a) = 10a^2(A-a)(1 + \sin(\pi/Aa))$ and $\mu(a) = \exp(-a)/(A-a)$, with $A = 1$; note that A has been rescaled to 1.

The following hypotheses on β, μ, and y_0 are in accordance with practical observations on biological populations.

(H1) $\beta \in L^\infty(Q_T)$, $\beta(a, t) \geq 0$ a.e. $(a, t) \in Q_T$.

(H2) $\mu \in L^1_{loc}([0, A) \times [0, T])$, $\mu(a, t) \geq 0$ a.e. $(a, t) \in Q_T$.

(H3) $y_0 \in L^1(0, A)$, $y_0(a) \geq 0$ a.e. $a \in (0, A)$.

$\qquad\qquad f \in L^1(Q_T)$, $f(a, t) \geq 0$ a.e. $(a, t) \in Q_T$.

By a solution to (4.1)–(4.3) we mean a function $y \in L^\infty(0, T; L^1(0, A))$, absolutely continuous along almost every characteristic line (of equation $a - t = k$, $(a, t) \in \overline{Q}_T$, $k \in \mathbb{R}$), that satisfies

$$\begin{cases} Dy(a, t) + \mu(a, t)y(a, t) = f(a, t) & \text{a.e. in } Q_T \\ y(0, t) = \displaystyle\int_0^A \beta(a, t)y(a, t)da & \text{a.e. } t \in (0, T) \\ y(a, 0) = y_0(a) & \text{a.e. } a \in (0, A). \end{cases} \qquad (4.4)$$

Because a solution y to (4.1)–(4.3) is absolutely continuous along almost every characteristic line, relations $(4.4)_{2,3}$ are meaningful; by $y(0, t)$ and $y(a, 0)$ we mean:

$$y(0, t) = \lim_{\varepsilon \to 0+} y(\varepsilon, t + \varepsilon) \quad \text{a.e. } t \in (0, T),$$

$$y(a, 0) = \lim_{\varepsilon \to 0+} y(a + \varepsilon, \varepsilon) \quad \text{a.e. } a \in (0, A).$$

Theorem 4.1. *Problem* (4.1)–(4.3) *admits a unique solution. The solution is nonnegative.*

Proof. Assume that y is a solution to (4.1)–(4.3). The definition of a solution allows us to conclude (by integrating along the characteristic lines) that

$$\begin{aligned} y(a, t) = {} & \exp\Big\{ -\int_0^a \mu(s, t - a + s)ds \Big\} b(t - a) \\ & + \int_0^a \exp\Big\{ -\int_s^a \mu(\tau, t - a + \tau)d\tau \Big\} f(s, t - a + s)ds \end{aligned} \qquad (4.5)$$

a.e. on $\{(a, t) \in Q_T;\ a < t\}$, and

$$\begin{aligned} y(a, t) = {} & \exp\Big\{ -\int_0^t \mu(a - t + s, s)ds \Big\} y_0(a - t) \\ & + \int_0^t \exp\Big\{ -\int_s^t \mu(a - t + \tau, \tau)d\tau \Big\} f(a - t + s, s)ds \end{aligned} \qquad (4.6)$$

a.e. on $\{(a, t) \in Q_T;\ a > t\}$.

Here we have denoted

$$b(t) = \int_0^A \beta(a,t)y(a,t)da \text{ a.e. } t \in (0,T). \tag{4.7}$$

The properties of β and y allow us to conclude (via (4.7)) that $b \in L^\infty(0,T)$. If we now assume that $b \in L^\infty(0,T)$, then y given by (4.5) and (4.6) is the unique solution to

$$\begin{cases} Dz(a,t) + \mu(a,t)z(a,t) = f(a,t), & (a,t) \in Q_T \\ z(0,t) = b(t), & t \in (0,T) \\ z(a,0) = y_0(a), & a \in (0,A). \end{cases} \tag{4.8}$$

Here, by a solution to (4.8) we mean a function $z \in L^\infty(0,T;L^1(0,A))$ absolutely continuous along almost every characteristic line, that satisfies

$$\begin{cases} Dz(a,t) + \mu(a,t)z(a,t) = f(a,t) & \text{a.e. in } Q_T \\ \lim_{\varepsilon \to 0+} z(\varepsilon, t + \varepsilon) = b(t) & \text{a.e. } t \in (0,T) \\ \lim_{\varepsilon \to 0+} z(a + \varepsilon, \varepsilon) = y_0(a) & \text{a.e. } a \in (0,A). \end{cases}$$

The solution y of (4.1)–(4.3) satisfies (4.5) and (4.6), where b is given by (4.7). So, we may infer that

$$\begin{aligned} b(t) &= \int_0^A \beta(a,t)y(a,t)da \\ &= \int_0^t \beta(a,t)\exp\Big\{ -\int_0^a \mu(s,t-a+s)ds \Big\}b(t-a)da \\ &\quad + \int_0^t \beta(a,t)\int_0^a \exp\Big\{ -\int_s^a \mu(\tau,t-a+\tau)d\tau \Big\}f(s,t-a+s)ds\, da \\ &\quad + \int_t^A \beta(a,t)\exp\Big\{ -\int_0^t \mu(a-t+s,s)ds \Big\}y_0(a-t)da \\ &\quad + \int_t^A \beta(a,t)\int_0^t \exp\Big\{ -\int_s^t \mu(a-t+\tau,\tau)d\tau \Big\}f(a-t+s,s)ds\, da, \end{aligned}$$

for almost all $0 < t < \min\{T,A\}$, and

$$\begin{aligned} b(t) &= \int_0^A \beta(a,t)y(a,t)da \\ &= \int_0^A \beta(a,t)\exp\Big\{ -\int_0^a \mu(s,t-a+s)ds \Big\}b(t-a)da \\ &\quad + \int_0^A \beta(a,t)\int_0^a \exp\Big\{ -\int_s^a \mu(\tau,t-a+\tau)d\tau \Big\}f(s,t-a+s)ds\, da, \end{aligned}$$

for almost all $\min\{T,A\} = A < t < T$ (if $A < T$).

In conclusion, b satisfies the following Volterra integral equation

$$b(t) = F(t) + \int_0^t K(t, t-s)b(s)ds \quad \text{a.e. } t \in (0, T) \tag{4.9}$$

(which is known as the renewal equation, or Lotka equation). Here

$$F(t) := \int_0^t \beta(a, t) \int_0^a \exp\Big\{ - \int_s^a \mu(\tau, t-a+\tau)d\tau \Big\} f(s, t-a+s)ds\, da$$
$$+ \int_t^A \beta(a, t) \int_0^t \exp\Big\{ - \int_s^t \mu(a-t+\tau, \tau)d\tau \Big\} f(a-t+s, s)ds\, da$$
$$+ \int_t^A \beta(a, t) \exp\Big\{ - \int_0^t \mu(a-t+s, s)ds \Big\} y_0(a-t)da$$

for almost all $0 < t < \min\{T, A\}$, and

$$F(t) := \int_0^A \beta(a, t) \int_0^a \exp\Big\{ - \int_s^a \mu(\tau, t-a+\tau)d\tau \Big\} f(s, t-a+s)ds\, da$$

for almost all $\min\{T, A\} = A < t < T$ (if $A < T$), and

$$K(t, a) := \begin{cases} \beta(a, t)e^{-\int_0^a \mu(s, t-a+s)ds} & \text{a.e. } (a, t) \in Q_T, \ a < t, \\ 0 & \text{elsewhere .} \end{cases}$$

Our hypotheses allow us to conclude that

$$\begin{cases} K \in L^\infty(Q_T), & K(t, a) \geq 0 \quad \text{a.e. } (a, t) \in Q_T \\ F \in L^\infty(0, T), & F(t) \geq 0 \quad\ \text{a.e. } t \in (0, T) . \end{cases} \tag{4.10}$$

We prove via Banach's fixed point theorem that the renewal equation has a unique solution $b \in L^\infty(0, T)$. For this we consider the following norm on $L^\infty(0, T)$.

$$\|w\| = \text{Ess sup}_{t \in (0, T)} (e^{-\gamma t} |w(t)|), \quad w \in L^\infty(0, T),$$

where $\gamma > 0$ is a constant indicated later; this norm is equivalent to the usual norm on $L^\infty(0, T)$.

Consider the operator

$$\mathcal{F} : L^\infty(0, T) \to L^\infty(0, T),$$

such that

$$(\mathcal{F}w)(t) = F(t) + \int_0^t K(t, t-s)w(s)ds \quad \text{a.e. } t \in (0, T) .$$

For any $b_1, b_2 \in L^\infty(0, T)$:

$$\|(\mathcal{F}b_1)(t) - (\mathcal{F}b_2)(t)\|$$

$$= \text{Ess sup}_{t \in (0,T)} \left(e^{-\gamma t} \left| \int_0^t K(t, t-s)(b_1 - b_2)(s)ds \right| \right)$$

$$\leq \text{Ess sup}_{t \in (0,T)} \left(e^{-\gamma t} \|K\|_{L^\infty(Q_T)} \int_0^t e^{\gamma s} e^{-\gamma s} |(b_1 - b_2)(s)| ds \right)$$

$$\leq \text{Ess sup}_{t \in (0,T)} \left(e^{-\gamma t} \|\beta\|_{L^\infty(Q_T)} \cdot \|b_1 - b_2\| \cdot \frac{1}{\gamma} e^{\gamma t} \right).$$

Hence for any $\gamma > \|\beta\|_{L^\infty(Q_T)}$, $\mathcal{F}$ is a contraction on $(L^\infty(0,T), \|\cdot\|)$. By Banach's fixed point theorem we conclude the existence of a unique solution $b \in L^\infty(0,T)$ to (4.9) and b is the limit in $(L^\infty(0,T), \|\cdot\|)$ of the sequence $\{b_n\}$, defined by

$$\begin{aligned} b_0(t) &= F(t) & \text{a.e. } t \in (0,T), \\ b_{n+1}(t) &= F(t) + \int_0^t K(t, t-s)b_n(s)ds & \text{a.e. } t \in (0,T), \quad n \in I\!N. \end{aligned} \tag{4.11}$$

By (4.10) we deduce that $b_n(t) \geq 0$ a.e. $t \in (0,T)$, for all $n \in I\!N$, and, in conclusion, b, the unique solution to (4.9), is nonnegative on $(0,T)$.

Consider now y given by (4.5) and (4.6), where b is the unique solution of (4.9). The hypotheses (H1)–(H3), (4.5) and (4.6) allow us to conclude that $y(a,t) \geq 0$ a.e. in Q_T and that $\mu y \in L^1(Q_T)$.

By again using (4.5) and (4.6), it follows that y is a solution to (4.1)–(4.3). The solution is unique because b, the solution of (4.9), is unique and is nonnegative because b, y_0, f are nonnegative.

Theorem 4.2. *If the mortality rate satisfies in addition*

(H4) $$\int_0^A \mu(a, t - A + a)da = +\infty \quad \text{a.e. } t \in (0,T),$$

where μ is extended by zero on $(0,A) \times (-\infty, 0)$, then the solution y to (4.1)–(4.3) satisfies

$$\lim_{\varepsilon \to 0+} y(A - \varepsilon, t - \varepsilon) = 0 \quad \text{a.e. } t \in (0,T),$$

that is, $y(A,t) = 0$ a.e. $t \in (0,T)$.

Proof. By (4.6) we have that, for almost any $t \in (0, \min\{T, A\})$, and for any $\varepsilon > 0$ sufficiently small,

$$\begin{aligned} y(A - \varepsilon, t - \varepsilon) &= \exp\Big\{ -\int_0^{t-\varepsilon} \mu(A - t + s, s)ds \Big\} y_0(A - t) \\ &\quad + \int_0^{t-\varepsilon} \exp\Big\{ -\int_s^{t-\varepsilon} \mu(A - t + \tau, \tau)d\tau \Big\} f(A - t + s, s)ds \\ &\longrightarrow \exp\Big\{ -\int_0^t \mu(A - t + s, s)ds \Big\} y_0(A - t) \\ &\quad + \int_0^t \exp\Big\{ -\int_s^t \mu(A - t + \tau, \tau)d\tau \Big\} f(A - t + s, s)ds = 0, \end{aligned}$$

as $\varepsilon \to 0+$ (we have used Lebesgue's theorem) because of (H4).

By (4.5) and (H4) we deduce that, for almost all $t \in (A, T)$ (if $A < T$) and for any $\varepsilon > 0$ sufficiently small,

$$
\begin{aligned}
y(A - \varepsilon, t - \varepsilon) = \exp\Big\{ &-\int_0^{A-\varepsilon} \mu(s, t - A + s)ds\Big\} b(t - A) \\
&+ \int_0^{A-\varepsilon} \exp\Big\{ -\int_s^{A-\varepsilon} \mu(\tau, t - \varepsilon + \tau)d\tau\Big\} f(s, t - A + s)ds \\
\longrightarrow \exp\Big\{ &-\int_0^A \mu(s, t - A + s)ds\Big\} b(t - A) \\
&+ \int_0^A \exp\Big\{ -\int_s^A \mu(\tau, t - A + \tau)d\tau\Big\} f(s, t - A + s)ds = 0,
\end{aligned}
$$

as $\varepsilon \to 0+$.

Remark 4.3. If y_0, β, μ, and f satisfy the more restrictive assumptions (as happens in real situations): $y_0 \in C([0, A])$, $\beta \in C(\overline{Q}_T)$, $\mu \in C([0, A) \times [0, T])$, and $f \in C(\overline{Q}_T)$, then it follows, in the same manner as in the proof of Theorem 4.1, that y, the solution to (4.1)–(4.3) is continuous on $D_1 = \{(a, t) \in \overline{Q}_T; \ a > t\}$ and on $D_2 = \{(a, t) \in \overline{Q}_T; \ a < t\}$.

Another extremely useful result is the following one.

Theorem 4.4. *Let y be the solution of (4.1)–(4.3).*

(i) If $f(a, t) > 0$ a.e. in Q_T, then $y(a, t) > 0$ a.e. in Q_T.
(ii) If $\beta_i, \mu_i, y_{0i}, f_i$ satisfy (H1)–(H3) ($i \in \{1, 2\}$) and

$$
\begin{aligned}
\beta_1(a, t) \geq \beta_2(a, t), \quad \mu_1(a, t) \leq \mu_2(a, t) \quad &\text{a.e. in } Q_T, \\
y_{01}(a) \geq y_{02}(a) \quad &\text{a.e. in } (0, A), \\
f_1(a, t) \geq f_2(a, t) \quad &\text{a.e. in } Q_T,
\end{aligned}
$$

then $y^1(a, t) \geq y^2(a, t)$ a.e. in Q_T, where y^i is the solution to (4.1)–(4.3), corresponding to $\beta := \beta_i$, $\mu := \mu_i$, $y_0 := y_{0i}$, $f := f_i$, $i \in \{1, 2\}$.
Moreover, if $f_n \to f$ in $L^1(Q_T)$, and f_n satisfy (H3), then

$$
y_n \to y
$$

in $L^\infty(0, T; L^1(0, A))$, where y_n are the solutions of (4.1)–(4.3) corresponding to $f := f_n$, respectively.

Proof. Theorem 4.1 allows us to conclude that the solution of (4.1)–(4.3) is given by (4.5) and (4.6), where b, the solution of (4.5), is nonnegative. If $f(a, t) > 0$ a.e. in Q_T, then (4.5) and (4.6) imply that $y(a, t) > 0$ a.e. in Q_T.

By (4.11) we have that the solution b of (4.5) can be obtained as the limit in $L^\infty(0, T)$ of the following iterative sequence,

$$\begin{cases} b_0^{\beta,\mu,y_0,f}(t) = F^{\beta,\mu,y_0,f}(t), \\ b_{n+1}^{\beta,\mu,y_0,f}(t) = F^{\beta,\mu,y_0,f}(t) + \int_0^t K^{\beta,\mu,y_0,f}(t,t-s) b_n^{\beta,\mu,y_0,f}(s) ds \end{cases} \quad (4.12)$$

a.e. $t \in (0,T)$, $n \in \mathbb{N}^*$, where $F^{\beta,\mu,y_0,f}$ and $K^{\beta,\mu,y_0,f}$ are f and K defined earlier (they both depend on β, μ, y_0, f and, we emphasize this).

If

$$\beta_1(a,t) \geq \beta_2(a,t), \quad \mu_1(a,t) \leq \mu_2(a,t) \quad \text{a.e. in } Q_T,$$

$$y_{01}(a) \geq y_{02}(a) \qquad\qquad\qquad\qquad \text{a.e. in } (0,A),$$

$$f_1(a,t) \geq f_2(a,t) \qquad\qquad\qquad\qquad \text{a.e. in } Q_T,$$

then

$$K^{\beta_1,\mu_1,y_{01},f_1}(t,a) \geq K^{\beta_2,\mu_2,y_{02},f_2}(t,a) \quad \text{a.e. in } Q_T,$$
$$F^{\beta_1,\mu_1,y_{01},f_1}(t) \geq F^{\beta_2,\mu_2,y_{02},f_2}(t) \quad \text{a.e. in } (0,T),$$

and by (4.12) we deduce $b^{\beta_2,\mu_2,y_{02},f_2}(t) \leq b^{\beta_1,\mu_1,y_{01},f_1}(t)$ a.e. in $(0,T)$. By (4.5) and (4.6) we may conclude that $y^1(a,t) \geq y^2(a,t)$ a.e. in Q_T.

Finally, if $f_n \to f$ in $L^1(Q_T)$, then we infer that $F_n \to F$ in $L^\infty(0,T)$ (where F_n is F corresponding to $f := f_n$). Indeed,

$$F_n(t) - F(t)$$

$$= \int_0^t \beta(a,t) \int_0^a e^{-\int_s^a \mu(\tau,t-a+\tau)d\tau}(f_n - f)(s,t-a+s) ds\, da$$

$$+ \int_t^A \beta(a,t) \int_0^t e^{-\int_s^t \mu(a-t+\tau,\tau)d\tau}(f_n - f)(a-t+s,s) ds\, da$$

for almost all $t \in (0,\min\{T,A\})$, and

$$F_n(t) - F(t) = \int_0^A \beta(a,t) \int_0^a e^{-\int_s^a \mu(\tau,t-a+\tau)d\tau}(f_n - f)(s,t-a+s) ds\, da$$

for almost all $A = \min\{T,A\} < t < T$ (if $A < T$). Consequently, we get

$$\|F_n - F\|_{L^\infty(0,T)} \leq \|\beta\|_{L^\infty(Q_T)} \cdot \|f_n - f\|_{L^1(Q_T)},$$

and this implies $F_n \to F$ in $L^\infty(0,T)$ as $n \to +\infty$.

By (4.9) and using Bellman's lemma (e.g., Appendix A.2 and [Ani00]) we get that $b_{f_n} \to b$ in $L^\infty(0,T)$, where b_{f_n} is the solution of (4.9) corresponding to $F := F_n$ (and consequently to $f := f_n$).

From (4.5) and (4.6) it follows that $y_n \to y$ in $L^\infty(0,T;L^1(0,A))$, and thus we get the final conclusion.

Remark 4.5. For any $f \in L^1(Q_T)$, $f(a,t) > 0$ a.e. in Q_T, the solution y of (4.1)–(4.3) satisfies

$$y(a,t) > 0 \text{ a.e. } (a,t) \in Q_T.$$

This shows that indeed the biological meaning of A is that of the maximal age of the population species.

Let us say a few words about the large time behavior of the solution of a linear age-dependent population dynamics in the case of time-independent vital rates and zero inflow. We deal with the following model,

$$\begin{cases} Dy(a,t) + \mu(a)y(a,t) = 0, & (a,t) \in Q \\ y(0,t) = \int_0^A \beta(a)y(a,t)da, & t \in (0,+\infty) \\ y(a,0) = y_0(a), & a \in (0,A), \end{cases} \quad (4.13)$$

where $Q = (0,A) \times (0,+\infty)$.

Assume that β, μ and y_0 satisfy

(A1) $\beta \in L^\infty(0,A)$, $\beta(a) \geq 0$ a.e. $a \in (0,A)$, $\beta \neq 0_{L^\infty(0,A)}$;
(A2) $\mu \in L^1_{loc}([0,A))$, $\mu(a) \geq 0$ a.e. $a \in (0,A)$,

$$\int_0^A \mu(a)da = +\infty,$$

(A3) $y_0 \in L^1(0,A)$, $y_0(a) > 0$ a.e. $a \in (0,A)$.

We denote

$$R = \int_0^A \beta(a)\, e^{-\int_0^a \mu(s)ds}\, da,$$

which is called the net reproduction rate.

Theorem 4.6. *If (A1)–(A3) hold, then*

$$\lim_{t\to\infty} \|y(t)\|_{L^\infty(0,A)} = 0, \text{ if } R < 1;$$
$$\lim_{t\to\infty} \|y(t)\|_{L^1(0,A)} = +\infty, \text{ if } R > 1;$$
$$\lim_{t\to\infty} \|y(t) - \tilde{y}\|_{L^\infty(0,A)} = 0, \text{ if } R = 1,$$

where y is the solution to (4.13) and $\tilde{y}$ is a nontrivial steady-state of $(4.13)_{1,2}$.

For the proof we recommend [Ani00] and [Ian95]. This result is emphasized by numerical tests.

Remark 4.7. Assume that (A1)–(A3) hold, and that $y_0 \in C([0,A])$. If the following compatibility condition is satisfied

(Ac) $$y_0(0) = \int_0^A \beta(a)y_0(a)da,$$

then it is possible to prove that b, the solution of (4.9), is continuous, $b(0) = y_0(0)$, and consequently by (4.5) and (4.6) we deduce that $y \in C([0, A] \times [0, +\infty))$ (see [Ani00]).

If the compatibility condition (Ac) is not satisfied, then $\{(a, a); a \in [0, A)\}$ is a discontinuity line for y, the solution to (4.13). This fact is also easily observed from the graph of y obtained with MATLAB.

The numerical solution

We now build a Euler-type approximation for system (4.1)–(4.3), with $f \equiv 0$. The approximation takes into account the fact that the approach in Theorem 4.1 is based on integration over the characteristic lines. We first introduce the grids

$$a_i = (i-1)h, \quad i = 1, 2, \ldots, M$$

and

$$t_j = (j-1)h, \quad j = 1, 2, \ldots, N,$$

where $h > 0$ is the grid step, and the right-hand side limits correspond to A and T, respectively, according to the formulae

$$M = 1 + A/h, \quad N = 1 + T/h.$$

For the approximates of functions we use the general notation

$$\varphi(a_i, t_j) \approx \varphi_i^{(j)}.$$

We discretize Equation (4.1) at (a_i, t_j), and get

$$\frac{y_i^{(j)} - y_{i-1}^{(j-1)}}{h} + \mu_i^{(j)} y_i^{(j)} = 0. \tag{4.14}$$

We have used here the backward finite-difference approximation for the first derivative. Let φ be a function of C^2-class. By Taylor's formula we have

$$\varphi(x-h) = \varphi(x) - h\varphi'(x) + \frac{h^2}{2}\varphi''(\xi),$$

for some $\xi \in (x-h, x)$. This yields

$$\varphi'(x) = \frac{\varphi(x) - \varphi(x-h)}{h} + O(h). \tag{4.15}$$

Note that, in order to obtain (4.14), formula (4.15) is applied along the diagonal of the combined grids over $[0, A] \times [0, T]$. From (4.14) we readily get

$$y_i^{(j)} = (1 + h\mu_i^{(j)})^{-1} y_{i-1}^{(j-1)}.$$

Because we take μ to be independent of t for our numerical tests, the above formula becomes

$$y_i^{(j)} = (1 + h\mu_i)^{-1} y_{i-1}^{(j-1)}. \tag{4.16}$$

To approximate the integral in Equation (4.2), we use the iterated trapezoidal formula, according to which

$$\int_c^d \varphi(x)dx = \frac{h}{2}[\varphi(c) + \varphi(d)] + h \sum_{i=2}^{n-1} \varphi(x_i) + O(h^2),$$

where $h > 0$ is the step of the grid

$$c = x_1 < x_2 < \cdots < x_n = d,$$

and we have assumed that $\varphi \in C^2([c,d])$.

We now reassemble the approximation methods to build a numerical algorithm for system (4.1)–(4.3). We put $z(i,j)$ here instead of $z(a_i, t_j)$.

Algorithm 4.1

/* Compute the solution on the first time level ($j = 1$) from the initial condition (4.3) */
    for $i = 1$ to M
        $y(i,1) = y0(a(i))$
    end–for
/* Compute the solution ascending with respect to time levels, using formula (4.16), and the trapezoidal rule */
    for $j = 2$ to N
        for $i = 2$ to M
            $y(i,j) = y(i-1, j-1)/(1 + h * \mu(i))$
        end–for
        for $i = 1$ to M
            $w(i) = \beta(i) * y(i,j)$
        end–for
        $y(1,j) = trapz(a, w)$
    end–for

Here $trapz$ is a function for numerical integration. Let us point out that $y(1,j)$ is unknown (not yet computed) in the loop

    for $i = 1$ to M
        $w(i) = \beta(i) * y(i,j)$
    end–for

Therefore $w(1)$ is also unknown at the moment of the call $trapz(a, w)$. This is a logical inference, but recall that for our numerical tests we take $\beta(1) = 0$ and hence $w(1) = 0$. Hence the statement $y(1,j) = trapz(a, w)$ works properly.

We now give the program corresponding to $\beta(a,t) = B \cdot a^2(A - a)(1 + \sin(\pi/Aa))$, with $B > 0$, $\mu(a,t) = \exp(-a)/(A - a)$, $y_0(a) = \exp(-a^2/2)$:

```
% file pop1.m
% Program to compute the density of population
% using the backward Euler method ( step is h )
clear
A = input('The maximal age is A : ') ;
T = input('The final time is T : ') ;
h = input('The discretization parameter is h : ') ;
ct = input('constant for beta : ') ; % constant B>0 for β
lw = input('LineWidth : ') ;
M = 1 + A/h
N = 1 + T/h
for i = 1:M
    a(i) = (i − 1)*h;
end
for j = 1:N
    t(j) = (j − 1)*h;
end
beta = ct*a.^2.*(A − a).*(1 + sin(pi/A*a)) ;
miu = exp(−a)./(A − a) ;
R = trapz(a,beta.*exp(−miu))
y = zeros(M,N);
for i = 1:M
    y(i,1) = exp(−0.5*a(i)^2);
end
for j = 2:N
    j
    for i = 2:M
        y(i,j) = y(i − 1,j − 1)/(1 + h*miu(i)) ;
    end
    for i = 1:M
        w(i) = y(i,j) ;
    end
    y(1,j) = trapz(a,beta.*w) ;
end
for i = 1:M
    for j = 1:N
        age(i,j) = a(i) ;
        time(i,j) = t(j) ;
    end
end
% make figures
K = max(max(y)) ;
meshz(age,time,y) ; % or mesh
axis([0 A 0 T 0 K])
xlabel('\bf Age a','FontSize',16)
```

```
ylabel('\bf Time t','FontSize',16)
zlabel('\bf y(a,t)','FontSize',16)
figure(2)
plot(a, beta,'LineWidth',lw) ; grid
title('\bf \beta and \mu','FontSize',16)
xlabel('\bf a','FontSize',16)
hold on
plot(a, miu,'r','LineWidth',lw) ;
text(0.6,4,'\bf \beta(a)','FontSize',16)
text(0.8,6,'\bf \mu(a)','FontSize',16)
hold off
```

The coordinates of the *text* statements depend on the graphs.

To obtain a more accurate approximation for Equation (4.1), we introduce the *centered finite-difference* scheme. Let now φ be a function of C^3-class. By Taylor's formula we have

$$\varphi(x + h) = \varphi(x) + h\varphi'(x) + \frac{h^2}{2}\varphi''(x) + \frac{h^3}{6}\varphi'''(\xi_1).$$

$$\varphi(x - h) = \varphi(x) - h\varphi'(x) + \frac{h^2}{2}\varphi''(x) + \frac{h^3}{6}\varphi'''(\xi_2).$$

By subtracting the two formulae above we obtain

$$\varphi'(x) = \frac{\varphi(x + h) - \varphi(x - h)}{2h} + O(h^2).$$

Using this, we get the corresponding finite-difference equation to approximate (4.1)

$$\frac{y_{i+1}^{(j+1)} - y_{i-1}^{(j-1)}}{2h} + \mu_i^{(j)} y_i^{(j)} = 0.$$

Taking into account that $\mu_i^{(j)} = \mu_i$, we readily get

$$y_{i+1}^{(j+1)} = y_{i-1}^{(j-1)} - 2h\mu_i y_i^{(j)}. \tag{4.17}$$

Now, formula (4.17) replaces (4.16) and we obtain the following algorithm.

Algorithm 4.2

```
/* Compute the solution on the first time level (j = 1) from the initial
condition (4.3) */
    for i = 1 to M
        y(i, 1) = y0(a(i))
    end-for
/* Compute the solution for the time level j = 2 using formula (4.16) and
the trapezoidal rule */
```

$j = 2$
$i = 2$ to M
    $y(i, j) = y(i - 1, j - 1)/(1 + h * \mu(i))$
end–for
for $i = 1$ to M
    $w(i) = \beta(i) * y(i, j)$
end–for
$y(1, j) = trapz(a, w)$
/* Compute the solution ascending with respect to time levels $j = 3, \ldots, N$
using formula (4.17) and the trapezoidal rule */
    for $j = 2$ to $N - 1$
        for $i = 2$ to $M - 1$
            $y(i + 1, j + 1) = y(i - 1, j - 1) - 2 * h * \mu(i) * y(i, j)$
        end–for
        $i = 2$
        $y(i, j + 1) = y(i - 1, j)/(1 + h * \mu(i - 1))$
        for $i = 1$ to M
            $w(i) = \beta(i) * y(i, j + 1)$
        end–for
        $y(1, j + 1) = trapz(a, w)$
    end–for

Let us point out that, for $j = 2$ and for $i = 2$, respectively, we have to use a one-step formula. For example, $y(:, 1)$ is computed from the initial condition (4.3), then $y(:, 2)$ is computed by formula (4.16), and finally $y(:, j)$ is computed by formula (4.17) for $j = 3, \ldots, N$.
We invite the reader to write a program for Algorithm 4.2, using *pop1.m* as the model.
For a numerical test, we take $A = 1$, $T = 1$, $h = 0.02$, $B = 10$, $lw = 4$. With both algorithms above we obtain accurate graphs for β and μ (Figure 4.1) and y (Figure 4.2). To obtain Figure 4.2, we have rotated the original figure produced by the program.

Remark 4.8. The discontinuity along the line of equation $a = t$ occurs because the initial datum $y_0 \in C([0, A])$ taken by us does not satisfy the compatibility condition

$$y_0(0) = \int_0^A \beta(a) y_0(a) da.$$

For more details see [Ani00].

The vectors *beta*, *miu*, and $y(:, 1)$ from the above programs, which correspond to the grid-values of the functions β, μ, and y_0, that is,

$$beta(i) = \beta(a_i), \quad miu(i) = \mu(a_i), \quad y(i, 1) = y_0(a_i), \quad i = 1, 2, \ldots, M,$$

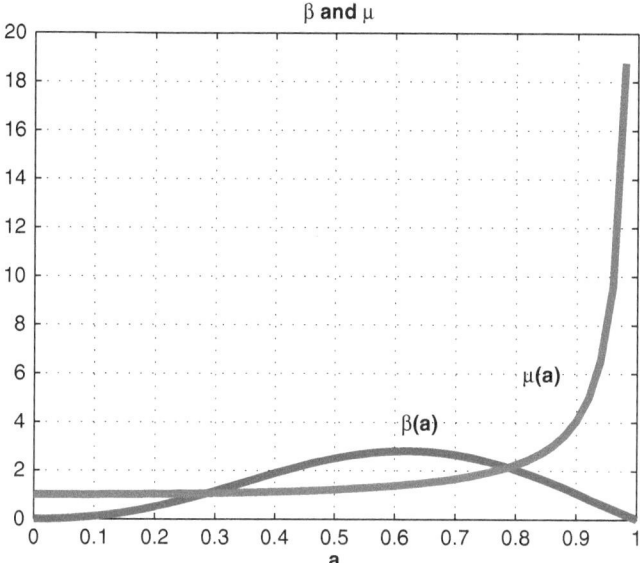

Fig. 4.1. The fertility and mortality rates

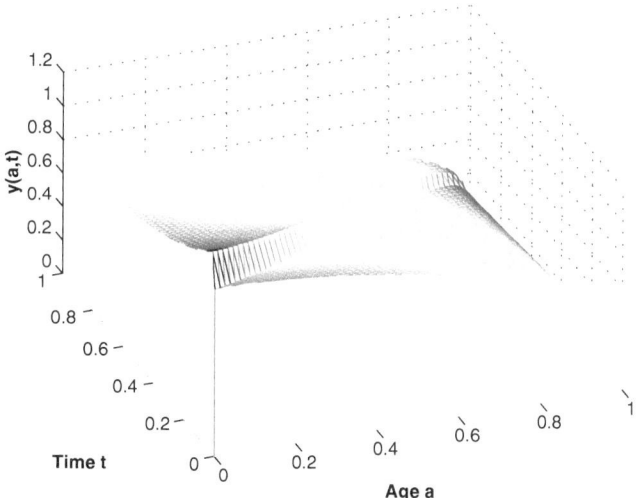

Fig. 4.2. The population distribution

can be loaded from files. Suppose that the values can be found in the files
"beta.txt" , "miu.txt", and "y0.txt", with the format one value per record.
Then we introduce the following program sequence.

load miu.txt
miu = miu' ; % transform into a row vector
load beta.txt
beta = beta' ; % transform into a row vector
load y0.txt
y0 = y0' ; % transform into a row vector

4.2 The optimal harvesting problem

This section concerns an optimal harvesting problem governed by a linear age-structured population dynamics. The corresponding PDE model was presented in the previous section. Our purpose is to prove the existence of an optimal control, obtain first-order necessary conditions of optimality, and use them to build a numerical algorithm. This algorithm leads to corresponding programs to approximate optimal control (and the optimal value of the cost functional).

The problem investigated here has great importance, both from an economical point of view (because it indicates the optimal strategy) as well as from a biological point of view.

We study an optimal harvesting problem; that is,

$$\text{Maximize} \int_0^T \int_0^A u(a,t)y^u(a,t)da\ dt, \qquad \textbf{(OHP)}$$

subject to $u \in K = \{w \in L^\infty(Q_T);\ 0 \le w(a,t) \le L \text{ a.e. in } Q_T\}$, where y^u is the solution to

$$\begin{cases} Dy(a,t) + \mu(a,t)y(a,t) = f(a,t) - u(a,t)y(a,t), & (a,t) \in Q_T \\ y(0,t) = \int_0^A \beta(a,t)y(a,t)da, & t \in (0,T) \qquad (4.18) \\ y(a,0) = y_0(a), & a \in (0,A). \end{cases}$$

Here $L > 0$, and K is the set of constrained controls. $u \in K$ is the control, or the harvesting effort. u plays the role of an additional mortality rate. The integral $\int_0^T \int_0^A u(a,t)y^u(a,t)da\ dt$ represents the total harvest.

Our goal is to find the control (the harvesting effort) that gives the maximal harvest: the optimal control (or the optimal harvesting effort).

The approach in this section is inspired by the paper [AA05].

For the sake of simplicity assume that assumptions (H1)–(H4) hold.

Because $\mu := \mu + u$ satisfies (H2) (for any $u \in K$), we conclude that system (4.18) has a unique solution y^u. This solution is nonnegative.

Existence of an optimal control

Theorem 4.9. *Problem (OHP) admits at least one optimal control.*

Proof. Define

$$\Psi(u) = \int_0^T \int_0^A u(a,t) y^u(a,t) da\ dt, \quad u \in K$$

and let

$$d = \sup_{u \in K} \Psi(u).$$

By Theorem 4.4 we get that

$$0 \le \Psi(u) \le L \int_0^T \int_0^A y^0(a,t) da\ dt < +\infty$$

(y^0 is the solution of (4.18) corresponding to $u \equiv 0$), for any $u \in K$, and consequently $d \in [0, +\infty)$.

Let $\{u_n\}_{n \in \mathbb{N}^*} \subset K$ be a sequence of controllers satisfying

$$d - \frac{1}{n} < \Psi(u_n) \le d.$$

The same Theorem 4.4 allows us to conclude that

$$0 \le y^{u_n}(a,t) \le y^0(a,t) \text{ a.e. in } Q_T$$

and so, on a subsequence, also denoted by $\{y^{u_n}\}$, we have

$$y^{u_n} \longrightarrow y^* \text{ weakly in } L^2(Q_T).$$

We recall now the following corollary of Mazur's theorem (see [BP86]).

Corollary 4.10. *(Mazur) Let $\{x_n\}_{n \in \mathbb{N}}$ be a sequence in a real Banach space X that is weakly convergent to $x \in X$. Then there exists a sequence $\{y_n\}_{n \in \mathbb{N}} \subset X$, $y_n \in \text{conv}\{x_k;\ k \ge n+1\}$, $n \in \mathbb{N}$, such that $\{y_n\}_{n \in \mathbb{N}}$ converges (strongly) to x.*

By using this corollary we obtain a sequence $\{\tilde{y}_n\}$ satisfying

$$\tilde{y}_n = \sum_{i=n+1}^{k_n} \lambda_i^n y^{u_i}, \quad \lambda_i^n \ge 0, \quad \sum_{i=n+1}^{k_n} \lambda_i^n = 1$$

($k_n \ge n+1$), and

$$\tilde{y}_n \to y^* \text{ in } L^2(Q_T).$$

Let the controls $\tilde{u}_n$ be defined as follows.

$$\tilde{u}_n(a,t) = \begin{cases} \dfrac{\sum_{i=n+1}^{k_n} \lambda_i^n y^{u_i}(a,t) u_i(a,t)}{\sum_{i=n+1}^{k_n} \lambda_i^n y^{u_i}(a,t)}, & \text{if } \sum_{i=n+1}^{k_n} \lambda_i^n y^{u_i}(a,t) \ne 0, \\ \\ 0 & \text{if } \sum_{i=n+1}^{k_n} \lambda_i^n y^{u_i}(a,t) = 0. \end{cases}$$

These controls satisfy $\tilde{u}_n \in K$ and

$$\tilde{y}_n(a,t) = y^{\tilde{u}_n}(a,t) \text{ a.e. in } Q_T.$$

We can take a subsequence (also denoted by $\{\tilde{u}_n\}$) such that

$$\begin{cases} \tilde{u}_n \to u^* & \text{weakly in } L^2(Q_T), \\ \Psi(\tilde{u}_n) = \displaystyle\sum_{i=n+1}^{k_n} \lambda_i^n \Psi(u_i) \to d & \text{as } n \to +\infty, \end{cases}$$

and we may infer that

$$d = \lim_{n \to +\infty} \Psi(\tilde{u}_n) = \int_0^T \int_0^A u^*(a,t) y^*(a,t) da\, dt. \qquad (4.19)$$

On the other hand the convergence $\tilde{u}_n \to u^*$ weakly in $L^2(Q_T)$ implies that $y^{\tilde{u}_n} \to y^{u^*}$ weakly in $L^2(Q_T)$ and this leads to

$$y^*(a,t) = y^{u^*}(a,t) \text{ a.e. in } Q_T$$

(because the weak limit is unique). By (4.19) we get that $d = \Psi(u^*)$ and consequently u^* is an optimal control for (OHP).

The maximum principle

Theorem 4.11. *Assume in addition that $f(a,t) > 0$ a.e. in Q_T. If u^* is an optimal control for Problem (OHP) and p is the solution of*

$$\begin{cases} Dp - \mu p = u^*(1+p) - \beta(a,t)p(0,t), & (a,t) \in Q_T \\ p(A,t) = 0, & t \in (0,T) \\ p(a,T) = 0, & a \in (0,A), \end{cases} \qquad (4.20)$$

then we have

$$u^*(a,t) = \begin{cases} 0 \text{ if } 1 + p(a,t) < 0 \\ L \text{ if } 1 + p(a,t) > 0. \end{cases} \qquad (4.21)$$

Proof. For any $v \in L^\infty(Q_T)$, such that $u^* + \varepsilon v \in K$ and for any $\varepsilon > 0$ sufficiently small, we get

$$\int_0^T \int_0^A u^* y^{u^*} da\, dt \geq \int_0^T \int_0^A (u^* + \varepsilon v) y^{u^* + \varepsilon v} da\, dt,$$

and this implies that

$$\int_0^T \int_0^A u^* \frac{y^{u^* + \varepsilon v} - y^{u^*}}{\varepsilon} da\, dt + \int_0^T \int_0^A v y^{u^* + \varepsilon v} da\, dt \leq 0. \qquad (4.22)$$

The convergence $y^{u^* + \varepsilon v} \to y^{u^*}$ in $L^\infty(0,T; L^1(0,A))$, as $\varepsilon \to 0+$ follows by Theorem 4.4.

Lemma 4.12. *The following convergence holds*

$$\frac{1}{\varepsilon}\left[y^{u^*+\varepsilon v} - y^{u^*}\right] \to z \quad \text{in } L^\infty(0,T;L^1(0,A)), \quad \text{as } \varepsilon \to 0+,$$

where z is the solution of

$$\begin{cases} Dz(a,t) + \mu(a,t)z(a,t) = -vy^{u^*} - u^*z, & (a,t) \in Q_T \\ z(0,t) = \int_0^A \beta(a,t)z(a,t)da, & t \in (0,T) \\ z(a,0) = 0, & a \in (0,A). \end{cases} \tag{4.23}$$

Proof of the lemma. Existence and uniqueness of z, the solution of (4.23), follows by Theorem 4.1. Denote by

$$w_\varepsilon(a,t) = \frac{1}{\varepsilon}\left[y^{u^*+\varepsilon v}(a,t) - y^{u^*}(a,t)\right] - z(a,t), \quad (a,t) \in Q_T.$$

It is obvious that w_ε is the solution of

$$\begin{cases} Dw(a,t) + \mu(a,t)w(a,t) = -u^*w - v\left[y^{u^*+\varepsilon v} - y^{u^*}\right], & (a,t) \in Q_T \\ w(0,t) = \int_0^A \beta(a,t)w(a,t)da, & t \in (0,T) \\ w(a,0) = 0, & a \in (0,A). \end{cases}$$

Because $y^{u^*+\varepsilon v} - y^{u^*} \longrightarrow 0$ in $L^\infty(0,T;L^1(0,A))$ as $\varepsilon \to 0+$ we infer via Theorem 4.4 that $w_\varepsilon \to 0$ in $L^\infty(0,T;L^1(0,A))$ as $\varepsilon \to 0+$ and this concludes the proof of the lemma.

Proof of Theorem 4.11, continued. By passing to the limit in (4.22) we may conclude

$$\int_0^T \int_0^A (u^*z + vy^{u^*})da\, dt \le 0. \tag{4.24}$$

By multiplying $(4.20)_1$ by z and integrating over $[0,T] \times [0,A]$ we get:

$$\int_0^T \int_0^A (Dp - \mu p)z\, da\, dt = \int_0^T \int_0^A [u^*(1+p)z - \beta(a,t)p(0,t)z(a,t)]da\, dt.$$

By using (4.20) and $(4.23)_2$, after some calculation, we obtain

$$-\int_0^T \int_0^A p(Dz + \mu z)da\, dt = \int_0^T \int_0^A u^*(1+p)z\, da\, dt. \tag{4.25}$$

If we use (4.25) we get that

$$\int_0^T \int_0^A pvy^{u^*}\, da\, dt = \int_0^T \int_0^A u^*z\, da\, dt,$$

and by (4.24) it follows that

$$\int_0^T \int_0^A v(a,t)(1+p(a,t))y^{u^*}(a,t)da\ dt \leq 0,$$

for any $v \in L^\infty(Q_T)$, such that $u^* + \varepsilon v \in K$, for any $\varepsilon > 0$ sufficiently small. This implies (along the same lines as in Chapter 2) that

$$u^*(a,t) = \begin{cases} 0 & \text{if } (1+p(a,t))y^{u^*}(a,t) < 0 \\ \\ L & \text{if } (1+p(a,t))y^{u^*}(a,t) > 0. \end{cases}$$

Because $f(a,t) > 0$ a.e. in Q_T, we conclude that $y^{u^*}(a,t) > 0$ a.e. in Q_T, and consequently (4.21) holds.

Remark 4.13. As a consequence of Theorem 4.11 we obtain that p (the solution of (4.20)) is a solution of

$$\begin{cases} Dp - \mu p = L(1+p)^+ - \beta(a,t)p(0,t), & (a,t) \in Q_T \\ p(A,t) = 0, & t \in (0,T) \\ p(a,T) = 0, & a \in (0,A). \end{cases} \tag{4.26}$$

For more details including uniqueness see [Ani00] and [AA05].

The uniqueness of the optimal control is clarified by the following result.

Remark 4.14. Let u^* be an optimal control for (OHP). Assume in addition that $f(a,t) > 0$ a.e. in Q_T, and that

(H5)
$$\begin{cases} \mu(a,t) > 0 \text{ a.e. in } Q_T, \\ \\ \text{and, for almost any } t \in (0,T),\ \dfrac{\beta}{\mu}(\cdot,t) \\ \text{is not a strictly positive constant on any subset} \\ \text{of positive measure.} \end{cases}$$

Under these additional conditions, equation $(4.20)_1$ implies that the set

$$\mathcal{D} = \{(a,t) \in Q_T;\ p(a,t) = -1\}$$

has Lebesgue measure zero (p is the solution of (4.20)), and u^* is a bang-bang control (u^* takes only a finite number of values, almost everywhere).

   In addition, the optimal control is unique.

Remark 4.15. If the hypotheses in the previous remark hold, then we may conclude that the adjoint state p (which is also the solution of (4.26)) does not depend on f or on y_0.

   Denote by u^* this optimal control.

The last remark allows us to formulate the following result for the general case $f(a,t) \geq 0$ a.e. in Q_T:

Theorem 4.16. *If (H1)–(H4) hold, then the control u^* is optimal also for Problem (OHP) corresponding to nonnegative inflow f.*

Proof. Consider $f_n \in L^1(Q_T)$, such that $f_n(a, t) > 0$ a.e. in Q_T and $f_n \to f$ in $L^1(Q_T)$ (as $n \to +\infty$). Denote by $\Psi_n(u)$ the cost function corresponding to $f := f_n$ in Problem (OHP), and by y_n^u the solution to (4.18) corresponding to u and $f := f_n$.

Theorem 4.4 allows us to conclude that for any $u \in K$,

$$y_n^u \to y^u$$

in $L^\infty(0, T; L^1(0, A))$, as $n \to +\infty$. This implies

$$\Psi_n(u) \to \Psi(u) \quad \text{as } n \to +\infty.$$

Because for any $u \in K$ we have $\Psi_n(u^*) \geq \Psi_n(u)$, and by using the last convergence, we conclude that u^* is an optimal control for (OHP). $\qquad \blacksquare$

Remark 4.17. If (H 1)–(H 4) hold, then (OHP) admits a unique optimal control u^*. In order to find u^* we first determine the solution p of (4.26). The optimal control is now given by (4.21).

The numerical solution

Theorem 4.11 and its consequence derived in formula (4.26) offer a simple way to get a numerical solution for the optimal control problem. Here is the algorithm.

- OH1: Compute p as the solution of Equation (4.26); that is,

$$\begin{cases} Dp - \mu p = L(1 + p)^+ - \beta(a)p(0, t), & (a, t) \in Q_T \\ p(A, t) = 0, & t \in (0, T) \\ p(a, T) = 0, & a \in (0, A). \end{cases}$$

- OH2: Compute the optimal control u^* according to formula (4.21).

Problem (OH1) is solved descending with respect to time levels.

Again using the backward finite-difference approximation for the differential operator we discretize the first equation (for p) by an explicit method

$$\frac{p_i^{(j)} - p_{i-1}^{(j-1)}}{h} - \mu_i p_i^{(j)} = L\Pi(p_i^{(j)}) - \beta_i p_1^{(j)}. \tag{4.27}$$

Here we have introduced the function

$$\Pi(p) = (1 + p)^+.$$

We therefore get

$$p_{i-1}^{(j-1)} = (1 - h\mu_i)p_i^{(j)} - h[L\Pi(p_i^{(j)}) - \beta_i p_1^{(j)}].$$

The corresponding loops to compute p are:

for $j = 1$ to N
    $p(M, j) = 0$
end–for
for $i = 1$ to M
    $p(i, N) = 0$
end–for
for $j = N$ down to 2
    for $i = 2$ to M
        $temp = h * (L * pp(p(i, j)) - \beta(i) * p(1, j))$
        $p(i - 1, j - 1) = (1 - h * \mu(i)) * p(i, j) - temp$
    end–for
end–for

Here pp is the positive part function denoted by Π in the mathematical formulae. The corresponding code follows.

```
function out = pp(p)
p1 = p + 1;
if (p1 >= 0)
    out = p1;
else
    out = 0;
end
```

Unfortunately, as concerns the numerical tests the explicit method has a drawback. We have considered in our tests the function

$$\mu(a) = \frac{e^{-a}}{A - a},$$

which (see the previous programs) reads

mu = exp $(-a)./(A - a)$;

It is quite clear that $mu(M)$ is not defined, and MATLAB gives to $mu(M)$ the value Inf, that is, $+\infty$. But $mu(i)$ is used inside the loop "for $i = 2$ to M" of the above algorithm. It follows that $mu(M)$ is used. Therefore we get NaN (Not a Number) for many values of p.

We therefore replace μ_i by μ_{i-1} and get the semi-implicit method

$$\frac{p_i^{(j)} - p_{i-1}^{(j-1)}}{h} - \mu_{i-1}p_{i-1}^{(j-1)} = L\Pi(p_i^{(j)}) - \beta_i p_1^{(j)},$$

which yields

$$p_{i-1}^{(j-1)} = (1 + h\mu_{i-1})^{-1}[p_i^{(j)} - h(L\Pi(p_i^{(j)}) - \beta_i p_1^{(j)})].$$

The corresponding algorithm is

Algorithm 4.3

```
for j = 1 to N
    p(M, j) = 0
end–for
for i = 1 to M
    p(i, N) = 0
end–for
for i = N downto 2
    for i = 2 to M
        temp = h * (L * pp(p(i, j)) − β(i) * p(1, j))
        p(i − 1, j − 1) = (p(i, j) − temp)/(1 + h * μ(i))
    end–for
end–for
for i = 1 to M
    for j = 1 to N
        if (1 + p(i, j) > 0)
            u(i, j) = L
        else
            u(i, j) = 0
        end–if
    end–for
end–for
```

The corresponding program is

```
% file ohp1.m
% Program for optimal harvesting – age dependent population
% (semi-implicit method)
clear
A = input('The maximal age is A = ') ;
T = input('The final time is T = ') ;
h = input('The discretization parameter is h = ') ;
L = input('L = ') ;
ct = input('constant for beta : ') ; % constant B>0 for β
M = 1 + A/h
N = 1 + T/h
N1 = N + 1 ;
for i = 1:M
    a(i) = (i − 1)*h ;
end
for j = 1:N
    t(j) = (j − 1)*h ;
```

```
end
beta = ct*a.∧2.*(A − a).*(1 + sin(pi/A*a)) ;
mu = exp(−a)./(A − a) ;
p = zeros(M,N) ;
u = zeros(M,N) ;
for j = 1:N
    p(M,j) = 0 ;
end
for i = 1:M
    p(i,N) = 0 ;
end
for k = 1:N − 1
    j = N1 − k
    for i = 2:M
        temp = h*(L*pp(p(i,j)) − beta(i)*p(1,j)) ;
        p(i − 1,j − 1) = (p(i,j) − temp)/(1 + h*mu(i − 1)) ;
    end
end
for i = 1:M
    for j = 1:N
        temp = 1 + p(i,j) ;
        if (temp > 0)
            u(i,j) = L ;
        end
    end
end

for i = 1:M
    for j = 1:N
        age(i,j) = a(i);
        time(i,j) = t(j);
    end
end
% make figures
K = max(max(u));
meshz(age,time,u);
axis([0 A 0 T 0 K]);
xlabel('\bf Age a','FontSize',16)
ylabel('\bf Time t','FontSize',16)
zlabel('\bf control u','FontSize',16)
figure(2)
meshz(age,time,p)
xlabel('\bf Age a','FontSize',16) ;
ylabel('\bf Time t','FontSize',16) ;
zlabel('\bf adjoint state p','FontSize',16) ;
```

```
tl = input('time level : ') ;
tval = t(tl)
for i = 1:M
    w(i) = u(i,tl) ;
end
figure(3)
plot(a,w,'ks') ; grid
axis([0 1 −2 12])
xlabel('\bf a','FontSize',16)
w1 = w' ;
save w1.txt w1 -ascii
```

Here pp is the function defined previously, and the array w is used to plot the graph of $a \mapsto u(a,t)$ for a given time level tl which corresponds to a time value t denoted $tval$ in the program.

We have tested the above program $ohp1.m$ for $A = 1$, $T = 1$, $L = 10$, $h = 0.005$, and $B = 10$ (the constant for function β). The corresponding optimal control u is given in Figure 4.3 and the adjoint state p in Figure 4.4.

Taking for the time level $tl = 81$ we have obtained the vector w corresponding to the time value $t = 0.4$. The section for the control u can be seen in Figure 4.5.

Another possibility is to use the implicit method to discretize the first equation from system (4.26). Therefore formula (4.27) is replaced by

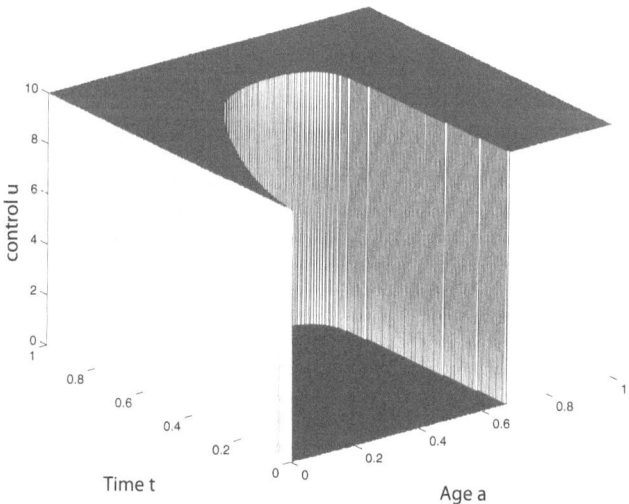

Fig. 4.3. The optimal control

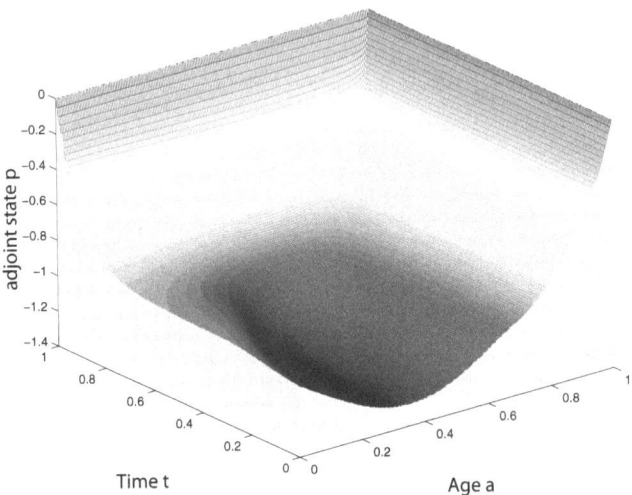

Fig. 4.4. The adjoint state p

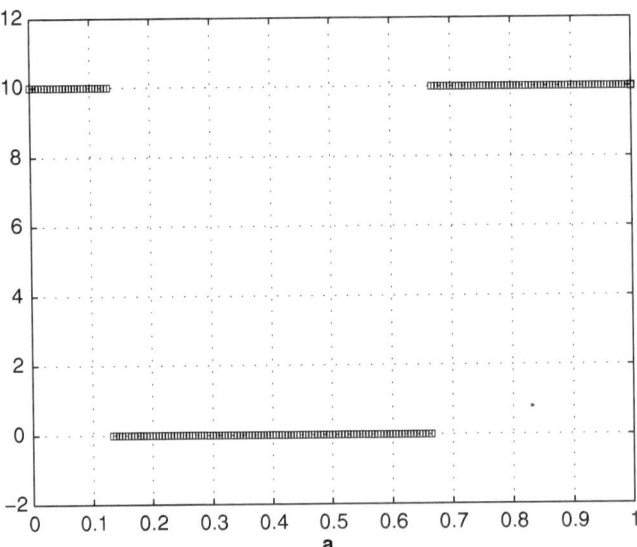

Fig. 4.5. The optimal control at $t = 0.4$

$$\frac{p_i^{(j)} - p_{i-1}^{(j-1)}}{h} - \mu_{i-1}p_{i-1}^{(j-1)} = L\Pi(p_{i-1}^{(j-1)}) - \beta_{i-1}p_1^{(j-1)},$$

and we get for $p_{i-1}^{(j-1)}$ the equation

$$a_1 p_{i-1}^{(j-1)} + a_2 \Pi(p_{i-1}^{(j-1)}) - a_3 = 0,$$

where
$$a_1 = 1 + h\mu_{i-1},$$
$$a_2 = hL,$$
$$a_3 = p_i^{(j)} + h\beta_{i-1}p_1^{(j-1)}.$$
It follows that $p_{i-1}^{(j-1)}$ is the solution of the following equation,

$$a_1 x + a_2 (x+1)^+ - a_3 = 0.$$

A simple calculation shows that the above equation admits two possible solutions, namely

$$x_1 = \frac{a_3 - a_2}{a_1 + a_2} \geq -1$$

and

$$x_2 = \frac{a_3}{a_1} < -1.$$

We choose one of them and the corresponding algorithm is

Algorithm 4.4

```
for j = 1 to N
    p(M, j) = 0
end–for
for i = 1 to M
    p(i, N) = 0
end–for
a2 = h * L;
for j = N downto 2
    for i = 2 to M
        a1 = 1 + h * μ(i − 1);
        a3 = p(i, j) + h * β(i − 1) * p(1, j − 1);
        x1 = (a3 − a2)/(a1 + a2);
        x2 = a3/a1;
        if (x1 < −1) and (x2 >= −1)
            error('NO ROOT');
        end–if
        if (x2 < −1)
            p(i − 1, j − 1) = x2;
        else
            p(i − 1, j − 1) = x1;
        end–if
    end–for
end–for
/* u is computed as in Algorithm 4.3 */
```

Let us remark that only the inside of the loop "for $i = 2$ to M" is modified with respect to Algorithm 4.3.

We invite the reader to write a program for Algorithm 4.4, using *ohp1.m* as the model. We suggest for a numerical test $A = 1$, $T = 1$, $L = 100$, $h = 0.01$, $B = 10$ (the multiplicative constant for function β) and $tl = 71$, that is, $t = 0.7$, for the time level.

A more complicated model is considered in the next section.

4.3 A logistic model with periodic vital rates

This section concerns an optimal harvesting problem for an age-structured population dynamics with logistic term and periodic vital rates. The starting point is the following nonlinear model of age-structured population dynamics

$$\begin{cases} Dy(a,t) + \mu(a,t)y + \mathcal{M}(t,Y(t))y = -u(t)y, & (a,t) \in Q \\ Y(t) = \displaystyle\int_0^A y(a,t)da, & t > 0 \\ y(0,t) = \displaystyle\int_0^A \beta(a,t)y(a,t)da, & t > 0 \\ y(a,0) = y_0(a), & a \in (0,A), \end{cases} \qquad (4.28)$$

where $Q = (0,A) \times (0,+\infty)$. The vital rates β (the fertility rate) and μ (the mortality rate), depending on age and time, are assumed to be T-periodic with respect to t. $Y(t)$ denotes the density of the total population at the moment t, and $\mathcal{M}(t,Y(t))$ stands for an additional mortality rate due to the overpopulation. The T-periodic function u is the harvesting effort (the control). Notice that system (4.28) is of IVP type. It has been proved in [AAA08] and [AAS09] that under certain natural hypotheses, the solution y^u to (4.28) satisfies

$$\lim_{t\to+\infty} \|y^u(t) - \tilde{y}^u(t)\|_{L^\infty(0,A)} = 0,$$

where $\tilde{y}^u$ is the maximal nonnegative solution to the following system, which is a periodic one

$$\begin{cases} Dy(a,t) + \mu(a,t)y + \mathcal{M}(t,Y(t))y = -u(t)y, & (a,t) \in Q \\ Y(t) = \displaystyle\int_0^A y(a,t)da, & t > 0, \\ y(0,t) = \displaystyle\int_0^A \beta(a,t)y(a,t)da, & t > 0, \\ y(a,t) = y(a,t+T), & (a,t) \in Q. \end{cases} \qquad (4.29)$$

In fact (4.29) has at most two nonnegative solutions ($y \equiv 0$ is of course a nonnegative solution for (4.29)).

In this context we consider the problem of finding the T-periodic harvesting effort u that leads to a maximal harvest on the time interval $[t, t+T]$ as $t \to +\infty$.

By taking into account that, for any positive initial datum y_0,

$$\int_t^{t+T} \int_0^A u(s)y^u(a,s)da\ ds \longrightarrow \int_0^T \int_0^A u(s)\tilde{y}^u(a,s)da\ ds,$$

as $t \to +\infty$ (the total harvest on $[t, t+T]$ for the population described by
(4.28) is tending to the total harvest on time intervals of length T for a periodic
population given by (4.29)), the problem may be reformulated as

$$\text{Maximize}\ \int_0^T \int_0^A u(s)\tilde{y}^u(a,s)da\ ds,\ \text{over all } u \in K, \qquad \textbf{(OH)}$$

where the set of controls K is now given by

$$K = \{v \in L^\infty(\mathbb{R}^+);\ 0 \le v(t) \le L,\ v(t) = v(t+T)\ \text{a.e. in } \mathbb{R}^+\},$$

and $L \in (0, +\infty)$.

Here are the hypotheses (see [AAA08] and [AAS09]):

(Hyp1)
$$\beta \in C(\mathbb{R}^+; L^\infty(0, A)),$$

$$\beta(a,t) \ge 0,\ \beta(a,t) = \beta(a,t+T)\ \text{a.e. } (a,t) \in Q,$$

there exist δ, $\tau > 0$ and $a_0 \in (0, A)$ such that $a_0 + T \le A$,
and $\beta(a, \tau) \ge \delta$ a.e. $a \in (a_0, a_0 + T)$,

This last assumption on β means that the fertile age-period for the population
species is longer than or equal to the period T.

(Hyp2)
$$\mu \in C(\mathbb{R}^+; L^\infty(0, \tilde{A}))\ \text{for any}\ \tilde{A} \in (0, A),$$

$$\mu(a,t) \ge 0,\ \mu(a,t) = \mu(a,t+T)\ \text{a.e. } (a,t) \in Q$$

(Hyp3) $\qquad\qquad y_0 \in L^\infty(0, A),\ y_0(a) > 0$ a.e. in $(0, A)$

(Hyp4) $\qquad\qquad \mathcal{M} : \mathbb{R}^+ \times \mathbb{R}^+ \to \mathbb{R}^+$ is a continuous function, continu-
ously differentiable with respect to the second variable and the derivative $\mathcal{M}_Y$
(with respect to the second variable) is positive on $\mathbb{R}^+ \times \mathbb{R}^+$. In addition

$$\mathcal{M}(t, Y) = \mathcal{M}(t+T, Y)\quad \text{for any } t,\ Y \in \mathbb{R}^+,$$

$$\mathcal{M}(t, 0) = 0\quad t \in \mathbb{R}^+,$$

$$\lim_{Y \to +\infty} \mathcal{M}(t, Y) = +\infty\ \text{uniformly with respect to } t.$$

By a solution to (4.28) we mean a function $y^u \in L^\infty(0, \tilde{T}; L^1(0, A))$ (for
any $\tilde{T} > 0$), absolutely continuous along almost every characteristic line (of
equation $a - t = k$, $(a,t) \in Q$, $k \in \mathbb{R}$), which satisfies

$$\begin{cases} Dy^u(a,t) = -\mu(a,t)y^u(a,t) - \mathcal{M}(t,Y^u(t))y^u(a,t) \\ \qquad\qquad -u(t)y^u(a,t) & \text{a.e. } (a,t) \in Q \\ Y^u(t) = \displaystyle\int_0^A y^u(a,t)da & \text{a.e. } t > 0 \\ y^u(0,t) = \displaystyle\int_0^A \beta(a,t)y^u(a,t)da & \text{a.e. } t > 0 \\ y^u(a,0) = y_0(a) & \text{a.e. } a \in (0,A). \end{cases}$$

By a solution to (4.29) we mean a function $\tilde{y}^u$, absolutely continuous along almost every characteristic line (of equation $a - t = k$, $(a,t) \in Q$, $k \in \mathbb{R}$), which satisfies

$$\begin{cases} D\tilde{y}^u(a,t) = -\mu(a,t)\tilde{y}^u(a,t) - \mathcal{M}(t,\tilde{Y}^u(t))\tilde{y}^u(a,t) \\ \qquad\qquad -u(t)\tilde{y}^u(a,t) & \text{a.e. } (a,t) \in Q \\ \tilde{Y}^u(t) = \displaystyle\int_0^A \tilde{y}^u(a,t)da & \text{a.e. } t > 0 \\ \tilde{y}^u(0,t) = \displaystyle\int_0^A \beta(a,t)\tilde{y}^u(a,t)da & \text{a.e. } t > 0 \\ \tilde{y}^u(a,t) = \tilde{y}^u(a,t+T) & \text{a.e. } (a,t) \in Q. \end{cases}$$

An ergodicity result established in [T84] implies the existence of a unique pair $(\alpha, y^*) \in \mathbb{R} \times C(\mathbb{R}^+; L^\infty(0,A))$ such that y^* is the nonnegative solution to

$$\begin{cases} Dy^*(a,t) + \mu(a,t)y^* + \alpha y^* = 0, & (a,t) \in Q \\ y^*(0,t) = \displaystyle\int_0^A \beta(a,t)y^*(a,t)da, & t > 0 \\ y^*(a,t) = y^*(a,t+T), & (a,t) \in Q, \end{cases} \qquad (4.30)$$

which satisfies

$$\text{Ess sup}\{y^*(a,t);\ (a,t) \in Q\} = 1.$$

Moreover, it follows that y^* is positive on $(0,A) \times (0,+\infty)$. The set of all solutions to (4.30) is a one-dimensional linear space.

Here is the asymptotic behavior result we have mentioned (see [Ani00] and [Ian95]).

Theorem 4.18. *For any $u \in K$, (4.28) has a unique solution y^u and*

$$\lim_{t \to +\infty} \|y^u(t) - \tilde{y}^u\|_{L^\infty(0,A)} = 0,$$

where $\tilde{y}^u$ is the maximal nonnegative solution to (4.29).
    Moreover,

(i) If $T\alpha > \int_0^T u(t)dt$, then $\tilde{y}^u$ is the unique nontrivial nonnegative solution to (4.29),

(ii) If $T\alpha \le \int_0^T u(t)dt$, then $\tilde{y}^u \equiv 0$ is the unique nonnegative solution to (4.29).

Remark 4.19. If $T\alpha > \int_0^T u(t)dt$ then $\tilde{y}^u(a,t) = c_0 y^*(a,t)h^u(t), (a,t) \in Q$, where $c_0 \in (0, +\infty)$ is an arbitrary constant and h^u is the unique nonnegative and nontrivial solution for

$$\begin{cases} h'(t) + \mathcal{M}(t, Y_0^*(t)h(t))h(t) - \alpha h(t) = -u(t)h(t), & t \in \mathbb{R}^+ \\ h(t) = h(t+T), & t \in \mathbb{R}^+, \end{cases} \quad (4.31)$$

where $Y_0^*(t) = c_0 \int_0^A y^*(a,t)da, \ t \geq 0$.

It is obvious that h^u depends on c_0 (via Y_0^*). However, $c_0 h^u$ is independent of c_0.

Remark 4.20. The result in Theorem 4.18 allows us to approximate the maximal nonnegative solution to (4.29) starting from the solution to (4.28) (see the algorithm (NSM) at the end of this section).

The optimal harvesting problem

By Theorem 4.18 we can see that for any $u \in K$:

$$\int_0^T \int_0^A u(t)\tilde{y}^u(t)da \ dt = \int_0^T u(t)Y_0^*(t)h^u(t)dt,$$

where h^u is the maximal nonnegative solution to (4.31) and so the optimal harvesting problem may be rewritten as

$$\text{Maximize } \int_0^T u(t)Y_0^*(t)h^u(t)dt, \quad \text{over all } u \in K. \quad \textbf{(OH)}$$

If $\alpha \leq 0$, then it follows by Theorem 4.18 that $h^u \equiv 0$, for any $u \in K$ and so Problem (OH) is trivial. This is the case when the population is going to extinction even without harvesting.

In what follows we treat the case

$$\alpha > 0 \ ,$$

when it is clear that if $\int_0^T u(t)dt$ is small, then $h^u(t) > 0$, for any $t \in [0, T]$.

We leave it to the reader to prove the existence of an optimal control as an exercise (see [AAS09]). However, we insist on first-order necessary optimality conditions.

Theorem 4.21. *Let (u^*, h^*) be an optimal pair for (OH). If q is the solution to*

$$\begin{cases} q'(t) - \mathcal{M}(t, Y_0^*(t)h^*(t))q(t) - \mathcal{M}_Y(t, Y_0^*(t)h^*(t))Y_0^*(t)h^*(t)q(t) \\ \qquad + \alpha q(t) = u^*(t)(Y_0^*(t) + q(t)), \\ q(t) = q(t+T), \quad t \geq 0, \end{cases} \quad (4.32)$$

then

$$u^*(t) = \begin{cases} 0, & if \ Y_0^*(t) + q(t) < 0 \\ L, & if \ Y_0^*(t) + q(t) > 0 \end{cases} \tag{4.33}$$

Proof. Remark that h^* is the positive solution to

$$\begin{cases} (h^*)'(t) = \gamma(t)h^*(t), & t > 0 \\ h^*(0) = h^*(T), \end{cases}$$

where $\gamma(t) = \alpha - u^*(t) - \mathcal{M}(t, Y_0^*(t)h^*(t))$ a.e. $t \in \mathbb{R}^+$. The T-periodicity of h^* implies that $\int_0^T \gamma(t)dt = 0$.

Our first goal is to prove that (4.32) has a unique solution q. Let us notice that (4.32) may be rewritten as

$$\begin{cases} q' = -\gamma(t)q + \mathcal{M}_Y(t, Y_0^*(t)h^*(t))Y_0^*(t)h^*(t)q(t) + u^*(t)Y_0^*(t), & t \in \mathbb{R}^+ \\ q(t) = q(t+T) & t \in \mathbb{R}^+. \end{cases}$$

A solution q satisfies

$$q(t) = q(0)\exp\{\int_0^t [-\gamma(s) + \mathcal{M}_Y(s, Y_0^*(s)h^*(s))Y_0^*(s)h^*(s)]ds\}$$
$$+ \int_0^t u^*(s)Y_0^*(s)\exp\{\int_s^t [-\gamma(\theta) + \mathcal{M}_Y(\theta, Y_0^*(\theta)h^*(\theta))Y_0^*(\theta)h^*(\theta)]d\theta\}ds,$$

for any $t \in \mathbb{R}^+$. Condition $q(0) = q(T)$ leads to

$$q(0) = q(0)\exp\{\int_0^T [-\gamma(s) + \mathcal{M}_Y(s, Y_0^*(s)h^*(s))Y_0^*(s)h^*(s)]ds\}$$
$$+ \int_0^T u^*(s)Y_0^*(s)\exp\{\int_s^T [-\gamma(\theta) + \mathcal{M}_Y(\theta, Y_0^*(\theta)h^*(\theta))Y_0^*(\theta)h^*(\theta)]d\theta\}ds,$$

for $t \in \mathbb{R}^+$, and consequently

$$q(0) = (1 - \exp\{\int_0^T [-\gamma(s) + \mathcal{M}_Y(s, Y_0^*(s)h^*(s))Y_0^*(s)h^*(s)]ds\})^{-1}$$
$$\cdot \int_0^T u^*(s)Y_0^*(s)\exp\{\int_s^T [-\gamma(\theta) + \mathcal{M}_Y(\theta, Y_0^*(\theta)h^*(\theta))Y_0^*(\theta)h^*(\theta)]d\theta\}ds < 0$$

(because of the assumptions on $\mathcal{M}$ and of the positivity of h^* and Y_0^*). The T-periodic function q, that we have obtained is the unique solution to (4.32).

Let $v \in L^\infty(\mathbb{R}^+)$ be an arbitrary T-periodic function such that $u^* + \varepsilon v \in K$, for any sufficiently small $\varepsilon > 0$. It is obvious that for sufficiently small $\varepsilon > 0$ we also have that $T\alpha > \int_0^T u^*(t)dt + \varepsilon \int_0^T v(t)dt$. Because u^* is an optimal control for (OH) we may infer that

$$\int_0^T u^*(t)Y_0^*(t)h^*(t)dt \geq \int_0^T (u^*(t) + \varepsilon v(t))Y_0^*(t)h^{u^*+\varepsilon v}(t)dt,$$

and

$$\cdot \int_0^T u^*(t)Y_0^*(t)\frac{h^{u^*+\varepsilon v}(t) - h^*(t)}{\varepsilon}dt + \int_0^T v(t)Y_0^*(t)h^{u^*+\varepsilon v}(t)dt \leq 0,$$

for any $\varepsilon > 0$ sufficiently small.

The following result is obtained as in the previous section.

Lemma 4.22. *The following convergences hold*

$$h^{u^*+\varepsilon v} \to h^* \qquad \text{in } C([0,T]),$$

$$\frac{1}{\varepsilon}[h^{u^*+\varepsilon v} - h^*] \to z \quad \text{in } C([0,T]),$$

as $\varepsilon \to 0+$, where z is the solution to the following problem,

$$\begin{cases} z' = \gamma(t)z - \mathcal{M}_Y(t, Y_0^*(t)h^*(t))Y_0^*(t)h^*(t)z(t) - v(t)h^*(t), & t \in I\!\!R^+ \\ z(t) = z(t+T), & t \in I\!\!R^+. \end{cases}$$

By passing to the limit ($\varepsilon \to 0+$) in the last inequality, and using the lemma we may conclude that

$$\int_0^T Y_0^*(t)[u^*(t)z(t) + v(t)h^*(t)]dt \leq 0.$$

By multiplying the first equation in (4.32) by z, and integrating over $[0,T]$ we get

$$\int_0^T q'(t)z(t)dt = \int_0^T z(t)[-\gamma(t)q(t)+\mathcal{M}_Y(t, Y_0^*(t)h^*(t))Y_0^*h^*q+u^*(t)Y_0^*(t)]dt.$$

In as much as

$$\int_0^T q'(t)z(t)dt = -\int_0^T q(t)z'(t)dt$$

$$= \int_0^T q(t)[-\gamma(t)z(t) + \mathcal{M}_Y(t, Y_0^*(t)h^*(t))Y_0^*(t)h^*(t)z(t) + v(t)h^*(t)]dt,$$

it follows that

$$\int_0^T u^*(t)Y_0^*(t)z(t)dt = \int_0^T v(t)h^*(t)q(t)dt.$$

This allows us to infer that

$$\int_0^T v(t)h^*(t)(Y_0^*(t) + q(t))dt \leq 0,$$

for any T-periodic $v \in L^\infty(I\!\!R^+)$ such that $u^* + \varepsilon v \in K$ and $T\alpha > \int_0^T u^*(t)dt + \varepsilon \int_0^T v(t)dt$, for any $\varepsilon > 0$ sufficiently small. The last relation implies (4.33).

Remark 4.23. By (4.32) and (4.33) we obtain that q is a solution to

$$
\begin{cases}
q'(t) - \mathcal{M}(t, Y_0^*(t)h^*(t))q(t) - \mathcal{M}_Y(t, Y_0^*(t)h^*(t))Y_0^*(t)h^*(t)q(t) \\
\quad + \alpha q(t) = L(Y_0^*(t) + q(t))^+, \\
q(t) = q(t + T), \quad t \in \mathbb{R}^+.
\end{cases}
\tag{4.34}
$$

Remark 4.24. If

$$
(Y_0^*)'(t) \neq \mathcal{M}(t, Y_0^*(t)h^*(t))Y_0^*(t) + \mathcal{M}_Y(t, Y_0^*(t)h^*(t))(Y_0^*(t))^2 h^*(t) - \alpha Y_0^*(t)
$$

a.e. in $\mathbb{R}^+$, then it follows by (4.34) that

$$
Y_0^*(t) + q(t) \neq 0 \ \text{a.e. in } (0, T).
$$

This implies, via (4.33), that

$$
u^*(t) = 0 \ \text{or } L \ \text{a.e. in } (0, T)
$$

(u^* is a bang-bang control).

Numerical algorithms

Next we develop a conceptual algorithm to approximate the solution of the optimal harvesting problem (OH). It is based on the optimality conditions in Theorem 4.21.

We first introduce an abstract Numerical Stabilization Method (NSM) for periodic differential equations. We consider the following abstract problem. Find $x \in X$ the unique solution of

$$
\begin{cases}
(Fx)(a, t) = z(a, t), & (a, t) \in Q \\
x(a, t) = x(a, t + T), & (a, t) \in Q,
\end{cases}
\tag{S}
$$

where $z \in Z$ is T-periodic with respect to t and $F : X \to Z$ is an appropriate operator between the appropriate spaces X and Z. Problem (S) is T-periodic with respect to t. As specified in the previous sections the solution of such a problem can be computed via the Initial-Value Problem (IVP)

$$
\begin{cases}
(Fx)(a, t) = z(a, t), & (a, t) \in Q \\
x(a, 0) = x_0(a), & a \in [0, A).
\end{cases}
\tag{S_0}
$$

The corresponding (NSM) is presented below.

Algorithm (NSM)

    Step 0 : Solve (S_0) for $t \in [0, T]$, and denote the numerical solution $x^{(0)}$; set $k := 1$;

    Step 1 : Solve

$$
\begin{cases}
(Fx)(a, t) = z(a, t), & a \in [0, A) \,, \ t \in [kT, (k+1)T] \\
x(a, kT) = x^{(k-1)}(a, kT), & a \in [0, A),
\end{cases}
\tag{S_k}
$$

and denote the numerical solution $x^{(k)}$;

Step 2 : The stopping criterion
$$\text{if } \|x^{(k)} - x^{(k-1)}\| < \varepsilon$$
then **stop** ($x^{(k)}$ is the solution)
else $k := k + 1$; go to Step 1.

In Step 2 above the norm should be an appropriate one, whereas $\varepsilon > 0$ is a given convergence parameter. For any time interval $[kT, (k+1)T]$ of length T we introduce a discretization grid with equidistant knots

$$kT = t_1 < t_2 < \cdots < t_N = (k+1)T$$

and we approximate the corresponding values of $x^{(k)} = (x_i^{(k)})_i$, $i = 1, 2, \ldots, N$, as $x_i^{(k)} \approx x^{(k)}(t_i)$, for $i = 1, 2, \ldots, N$. Of course every $x_i^{(k)}$ also needs a grid approximation with respect to a. The above norm can be a discrete one with respect to $x^{(k)} = (x_i^{(k)})_i$. When the (NSM) stops, $x^{(k)}$ is considered to be the T-periodic solution of Problem (S).

The biological system analyzed here has the stabilization property required by the (NSM). Independently of the initial-value function x_0 the solution x is stabilized after some time intervals of length T.

Now we present a Projected Gradient Method (PGM) for the optimal harvesting problem. By taking into account the control restrictions, we use Rosen's algorithm (e.g., [AN03, p. 44]). We also use Algorithm (NSM) above as a subroutine. To simplify the formulae we consider in the sequel $\mathcal{M}(t, Y) = Y$ and $c_0 = 1$. The corresponding (PGM) is presented below. Notice that it works for $\alpha > 0$.

Algorithm (PGM)

S0 : Compute the parameter α.

  S0.0 : Choose $y_0(a) > 0$ and solve the following system using Algorithm (NSM)

$$\begin{cases} Dy(a,t) + \mu(a,t)y + Y(t)y(a,t) = 0, & (a,t) \in Q \\ Y(t) = \displaystyle\int_0^A y(a,t)da, & t > 0 \\ y(0,t) = \displaystyle\int_0^A \beta(a,t)y(a,t)da, & t > 0 \\ y(a,t) = y(a,t+T), & (a,t) \in Q. \end{cases}$$

S0.1 : $\alpha = \dfrac{1}{T}\displaystyle\int_{kT}^{(k+1)T}\int_0^A y(a,t)da\, dt$,
  where k is obtained by Algorithm (NSM) in Step 2.

S0.2 : Set $j := 0$, $u^{(j)}(t) := u_0(t)$, where u_0 is given.

S1 : Compute $y^{(j)}$ the solution of the system

$$\begin{cases} Dy(a,t) + \mu(a,t)y + Y(t)y(a,t) = -u^{(j)}(t)y(a,t), & (a,t) \in Q \\ Y(t) = \int_0^A y(a,t)da, & t > 0 \\ y(0,t) = \int_0^A \beta(a,t)y(a,t)da, & t > 0 \\ y(a,t) = y(a,t+T), & (a,t) \in Q, \end{cases}$$

using Algorithm (NSM). Here $y^{(j)}$ corresponds to the solution $\tilde{y}^u$ of system (4.29).

S2 : Compute $h^{(j)}$ the solution of the system

$$\begin{cases} h'(t) + \left(\int_0^A y^{(j)}(a,t)da \right) h(t) - \alpha h(t) = -u^{(j)}(t)h(t), \, t \in \mathbb{R}^+ \\ h(t) = h(t+T), & t \in \mathbb{R}^+, \end{cases}$$

using Algorithm (NSM). Here $h^{(j)}$ corresponds to the solution of system (4.31).

S3 : Compute $q^{(j)}$.

S3.1 : $Y_0^{(j)}(t) = (h^{(j)}(t))^{-1} \int_0^A y^{(j)}(a,t)da.$

S3.2 : Compute $q^{(j)}$ the solution of the system

$$\begin{cases} q'(t) - 2 \left(\int_0^A y^{(j)}(a,t)da \right) q(t) + \alpha q(t) = u^{(j)}(t) \left(Y_0^{(j)}(t) + q(t) \right), & t \in \mathbb{R}^+ \\ q(t) = q(t+T), \, t \in \mathbb{R}^+, \end{cases}$$

computing first $q(0)$. Here $q^{(j)}$ corresponds to the solution of system (4.32);

S4 : Compute $w^{(j)}$ according to the formula

$$w^{(j)}(t) = \begin{cases} 0, & \text{if } Y_0^{(j)}(t) + q^{(j)}(t) < 0 \\ L, & \text{if } Y_0^{(j)}(t) + q^{(j)}(t) > 0. \end{cases}$$

Here $w^{(j)}$ is derived using u^* from formula (4.33).

S5 : Compute the new control $u^{(j+1)}$.
  S5.1 : Compute $\lambda_j \in [0,1]$ the solution of the maximization problem
  $$\max \{\Phi(\lambda u^{(j)} + (1-\lambda)w^{(j)}) \, ; \, \lambda \in [0,1]\},$$
where Φ is the cost functional corresponding to (OH)
  S5.2 : Compute $u^{(j+1)} = \lambda_j u^{(j)} + (1-\lambda_j)w^{(j)}.$

S6 : The stopping criterion.
  if $\|u^{(j+1)} - u^{(j)}\| < \varepsilon$

then **stop** ($u^{(j+1)}$ is the solution)
else $j := j + 1$; go to S1.

In S6 above the norm should be an appropriate one, whereas $\varepsilon > 0$ is a given convergence parameter.

Numerical experiments

We first point out that a convex combination of two bang-bang controls that take only the values 0 and L, does not necessarily take only these two values. This is why it is better to replace the convex combination $\lambda u^{(j)} + (1 - \lambda) w^{(j)}$ from Step 5.1 of Algorithm (PGM) using an idea due to K. Glashoff and E. Sachs (see [GS77] and for more details [AN03, pp.137–143]). The idea is to take a convex combination of the switching points of $u^{(j)}$ and of $w^{(j)}$ thus obtaining a system of switching points for the new bang-bang control.

We propose that reader write a program (based on Algorithms (NSM) and (PGM)) to approximate the optimal value of the total harvest. Use the following data.

$$\beta(a,t) = B \cdot a^2(1 - a)(1 + \sin(\pi a)) \left| \sin \frac{2\pi t}{T} \right|,$$

$$\mu(a,t) = \frac{e^{-4a}(2 + \cos \frac{2\pi t}{T})}{(1 - a)^{1.4}},$$

and $A = 1$, $T = 0.5$, $y_0(a) = 3$, $L = 1.5$, and $B \in \{50,\ 75,\ 100\}$.

Open problems

- Investigate the problem of maximizing the harvesting $\int_0^A u(a)y^u(a,t)da$ as $t \to +\infty$ when β and μ are time-independent and the control u belongs to

$$\mathcal{V} = \{v \in L^\infty(0, A);\ 0 \le v(a) \le L \text{ a.e. } a \in (0, A)\}.$$

- Investigate the problem of maximizing the harvesting

$$\int_t^{t+T} \int_0^A u(a,s)y^u(a,s)da\ ds \text{ as } t \to +\infty \text{ when the control } u \text{ belongs to}$$

$$\mathcal{V} = \{v \in L^\infty(Q);\ 0 \le v(a,t) \le L,\ v(a,t) = v(a, t + T) \text{ a.e. } (a, t) \in Q\}.$$

Bibliographical Notes and Remarks

There is an extensive literature devoted to the problem of optimal harvesting for age-dependent population dynamics with prescribed initial density of population. Here is a list of some of the most important results concerning this subject: [GM81], [B85], [Bar94], [AIK98], [Ani00], [HY05], and [FL06]. For the analysis and control of age-dependent population dynamics in a multi-layer environment we refer to [CIM05], and [IM09].

There are, however, only a few papers devoted to the important case of periodic age-structured population dynamics. We refer to [AIK98], [Ani00], and [LLW04] for linear models and to [AAA08] for nonlinear ones.

Numerical results for optimal harvesting problems can be found in [AA05] and [AAS09].

Exercises

4.1. Compute and plot the graph of the solution to the following problem.

$$
\begin{cases}
Dy(a,t) + \dfrac{a}{\pi - a} y(a,t) = t, & (a,t) \in (0,\pi) \times (0,1) \\[2mm]
y(0,t) = \displaystyle\int_0^\pi (\sin a) y(a,t) da, & t \in (0,1) \\[2mm]
y(a,0) = 1, & a \in (0,\pi).
\end{cases}
$$

4.2. Derive the first-order necessary optimality conditions for the following more general problem.

$$
\text{Maximize } \int_0^T \int_0^A u(a,t) g(a) y^u(a,t) da\, dt, \qquad \textbf{(OHP1)}
$$

subject to $u \in K = \{w \in L^\infty(Q_T);\ 0 \le w(a,t) \le L \text{ a.e. in } Q_T\}$, where y^u is the solution to (4.18). Here L is a positive constant.

We assume that g satisfies

$$
g \in C^1([0,A]), \quad g(a) > 0 \text{ for any } a \in [0,A].
$$

Here $g(a)$ is the weight (or cost) of an individual of age a. So,

$$
\int_0^T \int_0^A u(a,t) g(a) y^u(a,t) da\, dt
$$

gives the total weight (or cost) of the harvested population.

Hint. We denote $z^u(a,t) = g(a) y^u(a,t)$ and reformulate Problem (OHP1) with respect to u and z^u. Then we use the same approach as in Section 4.2.

4.3. Compute the optimal effort and the optimal cost of the harvested population for Problem (OHP1). Take for β, μ and y_0 as in the numerical tests in Section 4.2, and $g(a) = 1 + a$.

4.4. Derive the first-order necessary conditions for the following problem.

$$
\text{Maximize } \int_0^T \int_0^A u(a,t) y^u(a,t) da\, dt, \qquad \textbf{(OHP2)}
$$

subject to $u \in V = \{w \in L^\infty(Q_T);\ w_1(a,t) \le w(a,t) \le w_2(a,t) \text{ a.e. in } Q_T\}$, where y^u is the solution to (4.18). Here $w_1,\ w_2 \in L^\infty(Q_T)$, $0 \le w_1(a,t) \le w_2(a,t)$ a.e. in Q_T.

Hint. Let u^* be an optimal control for Problem (OHP2), and the adjoint state p the solution to

$$
\begin{cases}
Dp + \mu p = u^*(1+p) - \beta(a,t)p(0,t), & (a,t) \in Q_T \\
p(A,t) = 0, & t \in (0,T) \\
p(a,T) = 0, & a \in (0,A).
\end{cases}
$$

Then we have

$$
u^*(a,t) = \begin{cases}
w_1(a,t) & \text{if } 1 + p(a,t) < 0 \\
w_2(a,t) & \text{if } 1 + p(a,t) > 0.
\end{cases}
$$

5

Optimal control of diffusive models

Mathematical biology has its roots in population ecology, which treats the mathematical modeling of interacting species along the lines established by the mathematicians A. Lotka (1924) and V. Volterra (1926) in terms of nonlinear ordinary differential equations.

The goal of the models à la Lotka–Volterra is to offer a quantitative description of the evolution of the interacting populations in time. However, there are important aspects that cannot be neglected on the spatial structure of the relevant populations.

This chapter is devoted to the study of two optimal control problems related to diffusive models. The maximum principles are deduced, and numerical algorithms to approximate the optimal values of the cost functionals are indicated. Furthermore, at the end of this chapter an exercise concerning an age- and space-structured population dynamics is proposed.

5.1 Diffusion in mathematical models

After the pioneering works of R. A. Fisher [F37], A. N. Kolmogorov, I. G. Petrovskii, and N. S. Piskunov [KPP37], and J. Skellam [S51] (see also [Ske73] and [Aro85]) the mathematical modeling of spatially structured populations has been carefully analyzed, giving rise to a very flourishing literature on the so-called reaction–diffusion systems, in which diffusion of the relevant populations is added to the nonlinear dynamics of their interaction (see [Oku80] and [Mur89]).

We start here with a reminder of the two most important and universally accepted ways for introducing diffusion: one as a trivial consequence of a conservation law combined with the well-known Fick's law, and the other due to a suitable rescaling in time and space of the simple random walk.

Consider the diffusion of a population or any other substance, whose spatial density at spatial position x and time t is denoted by $y(x, t)$ (here $x \in \overline{\Omega} \subset I\!\!R^N$

S. Aniţa et al. *An Introduction to Optimal Control Problems in Life Sciences and Economics*, Modeling and Simulation in Science, Engineering and Technology, DOI 10.1007/978-0-8176-8098-5_5, © Springer Science+Business Media, LLC 2011

is the habitat, $N \in I\!\!N^*$ and $t \geq 0$); the function y is such that the population at time $t \geq 0$, in any region $V \subset \Omega$ (we take open and bounded subsets V with C^1-class boundary ∂V) is given by

$$Y_V(t) = \int_V y(x,t) dx. \tag{5.1}$$

Let V be an arbitrary subregion of Ω with the above-mentioned properties. We assume for the time being that y is smooth enough.

According to Fick's law "the flux of the population through $x \in \partial V$ is given by

$$J(x,t) = (\gamma(x,t)\nabla_x y(x,t)) \cdot \nu(x),$$

where $\nu(x)$ is the outward normal versor at x to ∂V, $\gamma(x,t)$ is the diffusion parameter, and $\nabla_x y = y_x$ denotes the gradient of y with respect to x."

If we denote by $f(x,t)$ any contribution to the population dynamics at $x \in V$, $t \geq 0$, a natural conservation law for the change of the population in region V is

$$Y_V'(t) = \int_{\partial V} J(x,t) d\sigma + \int_V f(x,t) dx.$$

By applying the divergence theorem and (5.1) we obtain

$$\int_V \left[\frac{\partial y}{\partial t}(x,t) - \mathrm{div}_x(\gamma(x,t)\nabla_x y(x,t)) - f(x,t) \right] dx = 0.$$

The subregion V is arbitrary, therefore the integrand must be zero, so that

$$\frac{\partial y}{\partial t}(x,t) - \mathrm{div}_x(\gamma(x,t)\nabla_x y(x,t)) - f(x,t) = 0. \tag{5.2}$$

In the case where γ is a constant (5.2) becomes

$$\frac{\partial y}{\partial t}(x,t) - \gamma \Delta y(x,t) = f(x,t), \tag{5.3}$$

where $\Delta = \Delta_x$ is the Laplacian with respect to x.

Fisher's equation is a particular case of (5.3) when we have a single population and $f(x,t)$ describes a birth-and-death process:

$$\frac{\partial y}{\partial t}(x,t) - \gamma \Delta y(x,t) = -\mu y(x,t) + \beta y(x,t) \left(1 - \frac{y(x,t)}{k} \right). \tag{5.4}$$

Here β is the natural fertility rate and μ is the natural death rate, whereas k is a positive constant. This is of course a logistic model (because $(\beta/k)y(x,t)$ is an additional mortality rate due to overpopulation, and is proportional to the population density).

Let us deduce now the diffusion equation in a different manner starting from a simple random walk.

A simple random walk is a Markov chain $(X_n)_{n \in \mathbb{N}}$ with a countable state space $E = \mathbb{Z}$, the set of all integers, and transition matrix

$$\text{prob}\,(X_{n+1} = k+1 \mid X_n = k) = p_{k,k+1} = p$$

$$\text{prob}\,(X_{n+1} = k-1 \mid X_n = k) = p_{k,k-1} = 1-p$$

$$\text{prob}\,(X_{n+1} = j \mid X_n = k) = p_{k,j} = 0,$$

for any $k \in E$ and $j \in E$, $j \neq k-1$, $j \neq k+1$, where $p \in (0,1)$. In the symmetric case $p = \frac{1}{2}$.

If we set $p_k(n) = \text{prob}(X_n = k)$, by the theorem of total probabilities, and the Markov property

$$\text{prob}(X_{n+1} = j \mid X_n = i, X_{n-1} = i_{n-1}, ..., X_0 = i_0)$$
$$= \text{prob}(X_{n+1} = j \mid X_n = i) = p_{i,j},$$

for any $i, j \in E$, we obtain

$$p_k(n) = \frac{1}{2} p_{k+1}(n) + \frac{1}{2} p_{k-1}(n);$$

that is,

$$p_k(n+1) - p_k(n) = \frac{1}{2}[p_{k+1}(n) - 2p_k(n) + p_{k-1}(n)]. \tag{5.5}$$

If we take $\Delta t \in \mathbb{R}_+^*$ as the unit time step, and $\Delta x \in \mathbb{R}_+^*$ as the unit space step, Equation (5.5) becomes

$$p_{k\Delta x}((n+1)\Delta t) - p_{k\Delta x}(n\Delta t) = \frac{1}{2}[p_{(k+1)\Delta x}(n\Delta t) - 2p_{k\Delta x}(n\Delta t) + p_{(k-1)\Delta x}(n\Delta t)].$$

Let $k\Delta x = x$, $n\Delta t = t$, and

$$p_{n\Delta x}(n\Delta t) := y(x,t);$$

we have, equivalently,

$$y(x, t+\Delta t) - y(x,t) = \frac{1}{2}[y(x+\Delta x, t) - 2y(x,t) + y(x-\Delta x, t)].$$

By a usual Taylor approximation, at the second order, the last equality becomes

$$y(x, t+\Delta t) - y(x,t) = \frac{1}{2}(\Delta x)^2 \frac{\partial^2 y}{\partial x^2}(x,t) + o((\Delta x)^2),$$

from which

$$\frac{y(x, t + \Delta t) - y(x,t)}{\Delta t} = \frac{1}{2} \frac{(\Delta x)^2}{\Delta t} \frac{\partial^2 y}{\partial x^2}(x,t) + \frac{o((\Delta x)^2)}{\Delta t}. \qquad (5.6)$$

If we now let $\Delta t \to 0$, $\Delta x \to 0$ in such a way that

$$\frac{(\Delta x)^2}{\Delta t} = 2\gamma, \text{ a constant,}$$

from (5.6) we obtain the usual diffusion equation with a constant diffusion coefficient

$$\frac{\partial y}{\partial t}(x,t) = \gamma \frac{\partial^2 y}{\partial x^2}(x,t), \quad x \in \mathbb{R}, \ t > 0.$$

If we impose as initial condition

$$y(\cdot, 0) = \delta_0$$

(the Dirac distribution at 0), from the theory of PDEs we obtain the fundamental solution

$$y(x,t) = \frac{1}{\sqrt{4\pi\gamma t}} \exp(-\frac{1}{4\gamma t}x^2), \quad x \in \mathbb{R}, \quad t > 0,$$

which is the probability density function of a Gaussian $N(0, 2\gamma t)$, which is the margin distribution of a Brownian motion at time t.

For a more extended discussion on this matter we refer the reader to [Ske73], [Oku80], and to the review article [O86].

As for the boundary conditions associated with (5.3), (5.4) or to other non-linear parabolic equations, recall that three of them are usually applied.

- The *Dirichlet condition* has the form:

$$y(x,t) = \varphi(x,t), \quad x \in \partial\Omega, \quad t \geq 0,$$

  where φ is a given function. When $\varphi \equiv 0$, this indicates (for biological populations) a completely inhospitable boundary $\partial\Omega$.
- The *Neumann condition* has the form:

$$\frac{\partial y}{\partial \nu}(x,t) = \varphi(x,t), \quad x \in \partial\Omega, \quad t \geq 0,$$

  where φ is a given function. For biological populations for instance, this condition means that the population flow through the boundary is φ. If $\varphi \equiv 0$, then there is no population flow through the boundary, and the population is isolated in the habitat Ω.
- The *Robin condition* has the form

$$\frac{\partial y}{\partial \nu}(x,t) + \alpha(x,t) = \varphi(x,t), \quad x \in \partial\Omega, \quad t \geq 0,$$

where α and φ are given functions. This condition usually follows (for populations) from the fact that the population flow through the boundary is proportional to the difference between the population density $y(x,t)$ inside Ω, and the population density $\tilde{y}(x,t)$ outside Ω:

$$\frac{\partial y}{\partial \nu}(x,t) = -\alpha(x,t)(y(x,t) - \tilde{y}(x,t)), \quad x \in \partial\Omega, \quad t \geq 0,$$

so that

$$\frac{\partial y}{\partial \nu}(x,t) + \alpha(x,t)y(x,t) = \alpha(x,t)\tilde{y}(x,t), \quad x \in \partial\Omega, \quad t \geq 0.$$

5.2 Optimal harvesting for Fisher's model

Consider the following Fisher's model describing the dynamics of a biological population that is free to move in an isolated habitat Ω.

$$\begin{cases} \dfrac{\partial y}{\partial t} - \gamma \Delta y = ry\left(1 - \dfrac{y}{k}\right) - m(x)u(x,t)y(x,t), & (x,t) \in Q_T \\[2mm] \dfrac{\partial y}{\partial \nu}(x,t) = 0, & (x,t) \in \Sigma_T \\[2mm] y(x,0) = y_0(x), & x \in \Omega. \end{cases} \quad (5.7)$$

Here Ω is a bounded open subset of $\mathbb{R}^N$ ($N \in \mathbb{N}^*$) with a C^1-class boundary. T, γ, r, k are positive constants, $Q_T = \Omega \times (0,T)$, $\Sigma_T = \partial\Omega \times (0,T)$, and $y_0 \in L^\infty(\Omega)$, $y_0(x) > 0$ a.e. $x \in \Omega$ (y_0 is the initial population density). This model derives from (5.4), when the natural fertility rate is greater than the natural mortality rate.

u is the harvesting effort acting only on a nonempty open subset $\omega \subset \Omega$, and m is the characteristic function of ω.
By convention

$$m(x)u(x,t) = \begin{cases} u(x,t), \ x \in \omega, & t \in (0,T) \\ 0, & x \in \Omega \setminus \overline{\omega}, \quad t \in (0,T). \end{cases}$$

The total harvest on the time interval $[0,T]$ is

$$\int_0^T \int_\omega u(x,t)y^u(x,t)dx \, dt,$$

where y^u is the solution to (5.7), and so a natural optimal control problem related to (5.7) is the following one.

$$\text{Maximize} \int_0^T \int_\omega u(x,t) y^u(x,t) dx \, dt, \qquad \textbf{(DP1)}$$

subject to $u \in K = \{w \in L^2(\omega \times (0,T)); \ 0 \le w(x,t) \le L \text{ a.e.}\} \ (L > 0)$, where y^u is the solution to (5.7).

For definitions and basic properties of solutions to parabolic equations we refer the reader to [Bar98].

If, for a $u \in K$, (5.7) admits a solution y^u, then the comparison principle for parabolic equations (see [PW84]) implies that this solution satisfies

$$0 < y^u(x,t) \le M = \|y_0\|_{L^\infty(\Omega)} e^{rT} \text{ a.e. } (x,t) \in Q_T.$$

By using Banach's fixed point result it follows that for any $u \in K$ the nonlinear problem (5.7) admits indeed a unique solution y^u, and this satisfies the above-mentioned double inequality.

Existence of an optimal control

Define

$$\Phi(u) = \int_0^T \int_\omega u(x,t) y^u(x,t) dx \, dt, \quad u \in K,$$

and let

$$d = \sup_{u \in K} \Phi(u).$$

By using the comparison principle for parabolic equations, we obtain that

$$0 < y^u(x,t) \le y^0(x,t) \text{ a.e. } (x,t) \in \Omega \times (0,T),$$

and so

$$0 \le \int_0^T \int_\omega u(x,t) y^u(x,t) dx \, dt \le L \int_0^T \int_\omega y^0(x,t) dx \, dt$$

As a consequence we derive that

$$d \in \mathbb{R}^+.$$

Let $\{u_n\}_{n \in \mathbb{N}^*} \subset K$ be a sequence of controllers satisfying

$$d - \frac{1}{n} < \Phi(u_n) \le d. \qquad (5.8)$$

Because $\{u_n\}_{n \in \mathbb{N}^*}$ is a bounded sequence in $L^2(\omega \times (0,T))$, it follows that there exists a subsequence, also denoted by $\{u_n\}_{n \in \mathbb{N}^*}$, such that

$$u_n \longrightarrow u^* \text{ weakly in } L^2(\omega \times (0,T)),$$

and
$$mu_n \longrightarrow mu^* \quad \text{weakly in } L^2(\omega \times (0,T)).$$

$u^* \in K$ because K is a closed convex subset of $L^2(\omega \times (0,T))$, and consequently it is weakly closed (see [Bre83]).

If we denote
$$a_n(x,t) = ry^{u_n}(x,t)\left(1 - \frac{y^{u_n}(x,t)}{k}\right) - m(x)u_n(x,t)y^{u_n}(x,t), \quad (x,t) \in Q_T,$$

then it is obvious that $\{a_n\}_{n \in \mathbb{N}^*}$ is bounded in $L^\infty(Q_T)$ (and in $L^2(Q_T)$), and y^{u_n} is the solution to

$$\begin{cases} \dfrac{\partial y}{\partial t} - \gamma \Delta y = a_n(x,t), & (x,t) \in Q_T \\[2mm] \dfrac{\partial y}{\partial \nu}(x,t) = 0, & (x,t) \in \Sigma_T \\[2mm] y(x,0) = y_0(x), & x \in \Omega. \end{cases}$$

The boundedness of $\{a_n\}_{n \in \mathbb{N}^*}$ implies (via the compactness result for parabolic equations; see [Bar98]) that there exists a subsequence such that

$$\begin{cases} a_{n_k} \longrightarrow a^* & \text{weakly in } L^2(Q_T) \\[2mm] y^{u_{n_k}} \longrightarrow y^* & \text{a.e. in } Q_T, \end{cases} \tag{5.9}$$

and that y^* is the solution to

$$\begin{cases} \dfrac{\partial y}{\partial t} - \gamma \Delta y = a^*(x,t), & (x,t) \in Q_T \\[2mm] \dfrac{\partial y}{\partial \nu}(x,t) = 0, & (x,t) \in \Sigma_T \\[2mm] y(x,0) = y_0(x), & x \in \Omega. \end{cases}$$

On the other hand, by (5.9), and by using the weak convergence of $\{u_n\}_{n \in \mathbb{N}^*}$ and the boundedness of $\{y^{u_n}\}_{n \in \mathbb{N}^*}$ in $L^\infty(Q_T)$, we may infer that

$$a_{n_k} \longrightarrow ry^*\left(1 - \frac{y^*}{k}\right) - mu^*y^* \quad \text{in } L^2(Q_T),$$

and consequently

$$a^* = ry^*\left(1 - \frac{y^*}{k}\right) - mu^*y^* \quad \text{in } L^2(Q_T).$$

In conclusion, y^* is the solution to (5.7) corresponding to $u := u^*$; that is, $y^* = y^{u^*}$.

If we pass to the limit in (5.8) we get that

$$d = \Phi(u^*),$$

and that u^* is an optimal control for problem (DP1).

The maximum principle

Problem (5.7) may be rewritten as an initial value problem in $L^2(\Omega)$:

$$\begin{cases} y'(t) = f(t, u(t), y(t)), & t \in (0, T) \\ y(0) = y_0, \end{cases}$$

where

$$f(t, u, y) = Ay + ry(1 - \frac{y}{k}) - muy.$$

Here A is a linear unbounded operator. In fact A is defined by

$$D(A) = \{w \in H^2(\Omega); \frac{\partial w}{\partial \nu} = 0 \text{ on } \partial\Omega\},$$

$$Ay = \gamma \Delta y, \quad y \in D(A).$$

For definitions and basic properties of the Sobolev spaces $H^k(\Omega)$ we refer the reader to [Ada75], and [Bre83]. The equation satisfied by the adjoint state p is suggested in Section 2.1:

$$p'(t) = -A^* p - rp + \frac{2r}{k} y^u p + mu(1 + p), \quad t \in (0, T).$$

Remark also that A is a self-adjoint operator, and so A^* may be replaced by A.

Actually, here is the result we prove.

Theorem 5.1. *If (u^*, y^{u^*}) is an optimal pair for (DP1), and if p is the solution to*

$$\begin{cases} \dfrac{\partial p}{\partial t} + \gamma \Delta p = -rp + \dfrac{2r}{k} y^{u^*} p + mu^*(1 + p), & (x, t) \in Q_T \\[2ex] \dfrac{\partial p}{\partial \nu}(x, t) = 0, & (x, t) \in \Sigma_T \\[2ex] p(x, T) = 0, & x \in \Omega, \end{cases} \qquad (5.10)$$

then we have

$$u^*(x, t) = \begin{cases} 0, & \text{if } 1 + p(x, t) < 0 \\[1ex] L, & \text{if } 1 + p(x, t) > 0 \end{cases} \qquad (5.11)$$

a.e. $(x, t) \in \omega \times (0, T)$.

Proof. Consider the set

$$V = \{w \in L^2(\omega \times (0,T)); \; u^* + \varepsilon w \in K \text{ for any } \varepsilon > 0 \text{ sufficiently small}\}.$$

For an arbitrary but fixed $v \in V$ we have that

$$\int_0^T \int_\omega u^*(x,t) y^{u^*}(x,t) dx \, dt \geq \int_0^T \int_\omega (u^*(x,t) + \varepsilon v(x,t)) y^{u^*+\varepsilon v}(x,t) dx \, dt,$$

and that

$$\int_0^T \int_\omega u^* \frac{y^{u^*+\varepsilon v} - y^{u^*}}{\varepsilon} dx \, dt + \int_0^T \int_\omega v y^{u^*+\varepsilon v} dx \, dt \leq 0, \qquad (5.12)$$

for any $\varepsilon > 0$ sufficiently small.

We postpone for the time being the proof of the following result.

Lemma 5.2. *The following convergences hold*

$$y^{u^*+\varepsilon v} \longrightarrow y^{u^*} \text{ in } L^\infty(Q_T),$$

$$\frac{1}{\varepsilon}\left[y^{u^*+\varepsilon v} - y^{u^*}\right] \longrightarrow z \text{ in } L^\infty(Q_T),$$

as $\varepsilon \longrightarrow 0+$, where z is the solution of

$$\begin{cases} \dfrac{\partial z}{\partial t} - \gamma \Delta z = rz - \dfrac{2r}{k} y^{u^*} z - mu^* z - mvy^{u^*}, & (x,t) \in Q_T \\[2mm] \dfrac{\partial z}{\partial \nu}(x,t) = 0, & (x,t) \in \Sigma_T \\[2mm] z(x,0) = 0, & x \in \Omega. \end{cases} \qquad (5.13)$$

Proof of theorem (continued). If we pass to the limit in (5.12) (and use Lebesgue's theorem and Lemma 5.2) we may conclude that

$$\int_0^T \int_\omega [u^*(x,t) z(x,t) + v(x,t) y^{u^*}(x,t)] dx \, dt \leq 0. \qquad (5.14)$$

We multiply the parabolic equation in (5.13) by p and we obtain that

$$\int_0^T \int_\Omega p\left[\frac{\partial z}{\partial t} - \gamma \Delta z\right] dx \, dt = \int_0^T \int_\Omega p\left[rz - \frac{2r}{k} y^{u^*} z - mu^* z - mvy^{u^*}\right] dx \, dt.$$

If we integrate by parts (with respect to t) and use Green's formula (with respect to x) we get that

$$-\int_0^T \int_\Omega z \left[\frac{\partial p}{\partial t} + \gamma \Delta p \right] dx \; dt = \int_0^T \int_\Omega p \left[rz - \frac{2r}{k} y^{u^*} z - mu^* z - mvy^{u^*} \right] dx \; dt.$$

Because p is a solution to system (5.10), we get that

$$-\int_0^T \int_\Omega z \left[-rp + \frac{2r}{k} y^{u^*} p + mu^*(1+p) \right] dx \; dt$$

$$= \int_0^T \int_\Omega p \left[rz - \frac{2r}{k} y^{u^*} z - mu^* z - mvy^{u^*} \right] dx \; dt,$$

and so, we may infer that

$$\int_0^T \int_\Omega m(x) u^*(x,t) z(x,t) dx \; dt = \int_0^T \int_\Omega m(x) v(x,t) y^{u^*}(x,t) p(x,t) dx \; dt.$$

This is equivalent to

$$\int_0^T \int_\omega u^*(x,t) z(x,t) dx \; dt = \int_0^T \int_\omega v(x,t) y^{u^*}(x,t) p(x,t) dx \; dt. \qquad (5.15)$$

Finally, by (5.14) and (5.15) we get that

$$\int_0^T \int_\omega v(x,t) y^{u^*}(x,t)(1 + p(x,t)) dx \; dt \le 0 \text{ for any } v \in V.$$

In the same manner as in Chapter 2 (and using the positivity of y^{u^*}) it follows that u^* satisfies (5.11), and that for any $u \in U = L^2(\omega \times (0,T))$ we can prove that

$$\Phi_u(u) = y^u(1 + p).$$

Let us prove now the first part of the lemma. The second part can be proved in the same manner.

For any $v \in V$, and for any $\varepsilon > 0$ sufficiently small, we have that

$$0 \le y^{u^*}(x,t), \quad y^{u^*+\varepsilon v} \le M \text{ a.e. } (x,t) \in Q_T.$$

We denote $w_\varepsilon = y^{u^*+\varepsilon v} - y^{u^*}$; w_ε is the solution to

$$\begin{cases} \dfrac{\partial w}{\partial t} - \gamma \Delta w = rw - \dfrac{r}{k} w(y^{u^*+\varepsilon v} + y^{u^*}) - muw - \varepsilon mvy^{u^*+\varepsilon v}, & (x,t) \in Q_T \\[2mm] \dfrac{\partial w}{\partial \nu}(x,t) = 0, & (x,t) \in \Sigma_T \\[2mm] w(x,0) = 0, & x \in \Omega. \end{cases}$$

If we again use the comparison result for parabolic equations, we obtain that

$$w_{1\varepsilon}(x,t) \le w(x,t) \le w_{2\varepsilon}(x,t) \quad \text{a.e. } (x,t) \in Q_T, \tag{5.16}$$

where $w_{1\varepsilon}$ is the solution to

$$
\begin{cases}
\dfrac{\partial w_1}{\partial t} - \gamma \Delta w_1 = \left(r + \dfrac{2rM}{k} + L\right) w_1 - \varepsilon M \|v\|_{L^\infty(Q_T)}, & (x,t) \in Q_T \\[2mm]
\dfrac{\partial w_1}{\partial \nu}(x,t) = 0, & (x,t) \in \Sigma_T \\[2mm]
w_1(x,0) = 0, & x \in \Omega,
\end{cases}
$$

and $w_{2\varepsilon}$ is the solution to

$$
\begin{cases}
\dfrac{\partial w_2}{\partial t} - \gamma \Delta w_2 = \left(r + \dfrac{2rM}{k} + L\right) w_2 + \varepsilon M \|v\|_{L^\infty(Q_T)}, & (x,t) \in Q_T \\[2mm]
\dfrac{\partial w_2}{\partial \nu}(x,t) = 0, & (x,t) \in \Sigma_T \\[2mm]
w_2(x,0) = 0, & x \in \Omega.
\end{cases}
$$

By direct verification we have that

$$w_{1\varepsilon}(x,t) = -\varepsilon M \|v\|_{L^\infty(Q_T)} \int_0^t e^{M_0(t-s)}\, ds \quad \text{a.e. } (x,t) \in Q_T,$$

and

$$w_{2\varepsilon}(x,t) = \varepsilon M \|v\|_{L^\infty(Q_T)} \int_0^t e^{M_0(t-s)}\, ds \quad \text{a.e. } (x,t) \in Q_T,$$

where $M_0 = r + (2rM)/(k) + L$.

It is obvious that

$$w_{1\varepsilon},\ w_{2\varepsilon} \longrightarrow 0 \text{ in } L^\infty(Q_T),$$

as $\varepsilon \to 0+$.

These convergences, and (5.16) imply that

$$w_\varepsilon \longrightarrow 0 \text{ in } L^\infty(Q_T),$$

as $\varepsilon \to 0+$, and this concludes the first part of the lemma.

For the second part take

$$l_\varepsilon = \frac{1}{\varepsilon}\left[y^{u^*+\varepsilon v} - y^{u^*}\right] - z$$

and prove in the same manner (by using comparison results for parabolic equations) that

$$l_\varepsilon \longrightarrow 0 \text{ in } L^\infty(Q_T),$$

as $\varepsilon \to 0+$ (which gives the conclusion of the second part of the lemma).

It is important to notice that the proof of the second convergence in the lemma is based on the first convergence.

If we are interested in investigating the problem without the logistic term, then we obtain:

$$\text{Maximize } \int_0^T \int_\omega u(x,t) y^u(x,t) dx \, dt, \qquad \text{(DP1')}$$

where $u \in K$, and y^u is the solution to

$$\begin{cases} \dfrac{\partial y}{\partial t} - \gamma \Delta y = ry - m(x)u(x,t)y(x,t), & (x,t) \in Q_T \\[2mm] \dfrac{\partial y}{\partial \nu}(x,t) = 0, & (x,t) \in \Sigma_T \\[2mm] y(x,0) = y_0(x), & x \in \Omega. \end{cases}$$

Existence of an optimal control follows in the same way.

Here are the first-order necessary conditions of optimality.

Theorem 5.3. *If (u^*, y^{u^*}) is an optimal pair for (DP1'), and if p is the solution of*

$$\begin{cases} \dfrac{\partial p}{\partial t} + \gamma \Delta p = -rp + u^*(1+p), & (x,t) \in Q_T \\[2mm] \dfrac{\partial p}{\partial \nu}(x,t) = 0, & (x,t) \in \Sigma_T \\[2mm] p(x,T) = 0, & x \in \Omega, \end{cases} \qquad (5.17)$$

then we have

$$u^*(a,t) = \begin{cases} 0 \text{ if } 1 + p(a,t) < 0 \\[2mm] L \text{ if } 1 + p(a,t) > 0 \end{cases} \qquad (5.18)$$

a.e. $(x,t) \in \omega \times (0,T)$.

By (5.17) and (5.18) we obtain in this case that p is the solution (which is unique) to

$$\begin{cases} \dfrac{\partial p}{\partial t} + \gamma \Delta p = -rp + mL(1+p)^+, & (x,t) \in Q_T \\[2ex] \dfrac{\partial p}{\partial \nu}(x,t) = 0, & (x,t) \in \Sigma_T \\[2ex] p(x,T) = 0, & x \in \Omega. \end{cases} \qquad (5.19)$$

If we approximate the solution p to this problem (which does not depend on u^* or on y^{u^*}), then by using (5.18) we immediately obtain u^*. Here is the algorithm.

- OP1: Compute p as the solution of Problem (5.19).
- OP2: Compute the optimal control by formula (5.18).

Problem (OP1) is solved by descending with respect to time levels. Of course the computed control is still an approximation due to the fact that Problem (OP1) is computed numerically. But the algorithm is quite simple.

By (5.17) and (5.18) we may also conclude that u^* is a bang-bang control. It also follows that the optimal control u^* does not depend explicitly on y_0.

Numerical algorithm for (DP1)

The last theorem and the form of Φ_u allow us to establish an algorithm to approximate the optimal value of the cost functional. We describe below a projected gradient-type algorithm (Uzawa's method).

Uzawa's method for (DP1)

S0: Choose $u^{(0)} \in K$;
$\quad\quad$ Set j:=0;
S1: Compute $y^{(j)}$, the solution to (5.7), with the input $u^{(j)}$:

$$\begin{cases} \dfrac{\partial y}{\partial t} - \gamma \Delta y = ry\left(1 - \dfrac{y}{k}\right) - mu^{(j)}y, & (x,t) \in Q_T \\[2ex] \dfrac{\partial y}{\partial \nu}(x,t) = 0, & (x,t) \in \Sigma_T \\[2ex] y(x,0) = y_0(x), & x \in \Omega. \end{cases}$$

S2: Compute $p^{(j)}$, the solution to (5.10), with the input $y^{(j)}$, $u^{(j)}$:

$$\begin{cases} \dfrac{\partial p}{\partial t} + \gamma \Delta p = -rp + \dfrac{2r}{k} y^{(j)} p + m u^{(j)} (1 + p), & (x,t) \in Q_T \\[3mm] \dfrac{\partial p}{\partial \nu}(x,t) = 0, & (x,t) \in \Sigma_T \\[3mm] p(x,T) = 0, & x \in \Omega. \end{cases}$$

S3: Compute the gradient direction $w^{(j)}$:

$$w^{(j)} := \Phi_u(u^{(j)}) = y^{(j)}(1 + p^{(j)}).$$

S4: (The stopping criterion)
   If $\|w^{(j)}\| < \varepsilon$
      then STOP ($u^{(j)}$ is the approximating control)
      else go to S5.

S5: Compute the steplength ρ_j such that

$$\Phi(P_K(u^{(j)} + \rho_j w^{(j)})) = \max_{\rho \geq 0} \{\Phi(P_K(u^{(j)} + \rho w^{(j)}))\}.$$

S6: Compute the new control

$$u^{(j+1)} := P_K(u^{(j)} + \rho_j w^{(j)}).$$

$j := j + 1$; go to S1.

Another possible choice is to use formula (5.11), and to develop a Rosen-type projected gradient method.

Rosen's method for (DP1)

S0: Choose $u^{(0)} \in K$;
         Set j:=0;
S1: Compute $y^{(j)}$ the solution to (5.7) with the input $u^{(j)}$:

$$\begin{cases} \dfrac{\partial y}{\partial t} - \gamma \Delta y = ry\left(1 - \dfrac{y}{k}\right) - m u^{(j)} y, & (x,t) \in Q_T \\[3mm] \dfrac{\partial y}{\partial \nu}(x,t) = 0, & (x,t) \in \Sigma_T \\[3mm] y(x,0) = y_0(x), & x \in \Omega. \end{cases}$$

S2: Compute $p^{(j)}$ the solution to (5.10) with the input $y^{(j)}$, $u^{(j)}$:

$$
\begin{cases}
\dfrac{\partial p}{\partial t} + \gamma \Delta p = -rp + \dfrac{2r}{k} y^{(j)} p + m u^{(j)} (1 + p), & (x,t) \in Q_T \\[3mm]
\dfrac{\partial p}{\partial \nu}(x,t) = 0, & (x,t) \in \Sigma_T \\[3mm]
p(x,T) = 0, & x \in \Omega.
\end{cases}
$$

S3: Compute $v^{(j)}$ according to the formula

$$
v^{(j)}(x,t) =
\begin{cases}
0 \ \text{if } 1 + p^{(j)}(x,t) < 0 \\[3mm]
L \ \text{if } 1 + p^{(j)}(x,t) \geq 0,
\end{cases}
$$

for $(x,t) \in \omega \times (0,T)$. Here $v^{(j)}$ is derived using u^* from formula (5.11).

S4: Compute $\lambda_j \in [0,1]$, the solution of the maximization problem

$$
\max_{\lambda \in [0,1]} \Phi(\lambda u^{(j)} + (1 - \lambda) v^{(j)}).
$$

S5: Compute the new control $u^{(j+1)}$ by

$$
u^{(j+1)} = \lambda_j u^{(j)} + (1 - \lambda_j) v^{(j)}.
$$

S6: (The stopping criterion)
  if $\|u^{(j+1)} - u^{(j)}\| < \varepsilon$
    then STOP ($u^{(j+1)}$ is the approximating control)
    else $j := j + 1$; go to S1.

We recall that a convex combination of two bang-bang controls that take only the values 0 and L, does not necessarily take only these two values. This is why it is better to replace the convex combination $\lambda u^{(j)} + (1 - \lambda) v^{(j)}$ from S4 and S5 by a convex combination of the switching points of $u^{(j)}$ and of $v^{(j)}$ as discussed in Section 4.3.

As concerns the use of MATLAB$^®$ we say that for $\Omega \subset \mathbb{R}$ one can apply finite-difference schemes for the approximation of the PDEs in S1 and S2 and write a corresponding program. For $\Omega \subset \mathbb{R}^2$ the Finite Element Method (FEM) can be used and therefore the Partial Differential Equation Toolbox^{TM} of MATLAB.

5.3 A working example: Control of a reaction–diffusion system

The model proposed for investigation in this section is related to the predator–prey model studied in Section 2.3.

Assume that the two populations are free to diffuse in the same habitat Ω.

A subsequent departure from various reaction–diffusion systems arising in population dynamics or epidemiology – and other application fields as well – lies in one of the reaction terms that is nonlocal in essence and involves an integral term. Going back to our motivating problem the biomass of captured and eaten prey at x in the habitat Ω is thereafter spatially distributed over the whole range occupied by the predator species, Ω. This produces local/nonlocal interspecific interactions between the two species at the predator level; that is, the functional response to predation is local whereas the numerical response to predation is nonlocal and distributed over Ω; see also [CW97], [G00], [GVA06], and [AFL09].

This is again a departure from most standard predator–prey models, see [Mur89]. From a phenomenological point of view we have chosen to introduce a rather generic integral kernel term to model the spatial distribution of biomass. For nonlocal diffusive epidemic models see [C84].

Let us now derive the mathematical problem with which we are working. Let $y_1(x,t)$ be the density at position x and time t of a prey species distributed over a spatial domain $\Omega \subset \mathbb{R}^N$, $N = 1$, 2, or 3, and assume its spatiotemporal dynamics is governed by the equation

$$\frac{\partial y_1}{\partial t} - d_1 \Delta y_1 = r_1 y_1, \quad x \in \Omega, \quad t \in (0, T),$$

wherein $r_1 > 0$ is the natural growth rate, $d_1 > 0$ is the diffusion coefficient, and $T > 0$. Let $y_2(x,t)$ be the density at position x and time t of a predator species distributed over the same habitat; in the absence of the aforementioned prey – assumed to be its unique resource – the predator population will decay at an exponential rate $r_2 > 0$ and its spatiotemporal dynamics is governed by a basic linear model,

$$\frac{\partial y_2}{\partial t} - d_2 \Delta y_2 = -r_2 y_2, \qquad x \in \Omega, \qquad t \in (0, T),$$

where $d_2 > 0$ is the diffusion coefficient.

When both populations are present, predation occurs on Ω; assume that the functional is of Lotka–Volterra type (see [Mur89]). The prey dynamics is modified by predation and reads

$$\frac{\partial y_1}{\partial t} - d_1 \Delta y_1 = r_1 y_1 - \mu_1 u(x,t) y_1 y_2, \quad x \in \Omega, \quad t \in (0, T), \quad (5.20)$$

where $\mu_1 > 0$ and $1 - u(x,t)$ is the segregation of the two populations. u is in fact the control.

Prey captured and eaten at time $t > 0$ and location $x' \in \Omega$ are transformed into biomass via a conversion factor yielding a numerical response to predation $\mu_2 u(x',t) y_1(x',t) y_2(x',t)$ ($\mu_2 > 0$ is a constant). We assume that this quantity is distributed over Ω via a generic nonnegative kernel $\ell(x, x')$ ($\ell \in L^\infty(\Omega \times \Omega)$, $\ell(x, x') \geq 0$ a.e. $(x, x') \in \Omega \times \Omega$) so that $\ell(x, x')\mu_2 u(x', t)y_1(x', t)y_2(x', t)$ is the biomass distributed at $x \in \Omega$ resulting from predation at $x' \in \Omega$. Biomass conservation implies a consistency condition must hold; $\int_\Omega \ell(x, x')dx = 1$, a.e. $x' \in \Omega$.

In this setting the predator dynamics reads

$$\frac{\partial y_2}{\partial t} - d_2 \Delta y_2 = -r_2 y_2 + \mu_2 \int_\Omega \ell(x, x')u(x', t)y_1(x', t)y_2(x', t)dx', \quad (5.21)$$

$x \in \Omega$, $t \in (0, T)$.

To complete our model boundary conditions must be imposed on both species. We choose no-flux boundary conditions corresponding to isolated populations:

$$\frac{\partial y_1}{\partial \nu}(x, t) = \frac{\partial y_2}{\partial \nu}(x, t) = 0, \quad x \in \partial\Omega, \quad t \in (0, T). \quad (5.22)$$

The last nonnegative and bounded initial conditions are prescribed at time $t = 0$:

$$\begin{cases} y_1(x, 0) = y_{01}(x), & x \in \Omega \\[2mm] y_2(x, 0) = y_{02}(x), & x \in \Omega. \end{cases} \quad (5.23)$$

Then (5.20)–(5.23) is a basic model for our controlled predator–prey system.

Assume that

$$y_{01}, \ y_{02} \in L^\infty(\Omega),$$

and

$$y_{01}(x) > 0, \quad y_{02}(x) > 0 \quad \text{a.e. } x \in \Omega.$$

We are interested in maximizing the total number of individuals of both populations at moment $T > 0$. This problem is related to (P3) in Section 2.3.

The problem may be reformulated:

$$\text{Maximize } \Psi(u) = \int_\Omega [y_1^u(x, T) + y_2^u(x, T)]dx, \quad \textbf{(DP2)}$$

subject to $u \in L^2(\Omega \times (0, T))$, $0 \leq u(x, t) \leq 1$ a.e. $t \in (0, T)$, where (y_1^u, y_2^u) is the solution to (5.20)–(5.23).

We propose that the reader proves that for any such u problem (5.20) and (5.21) admits a unique solution, and that the following result holds.

Theorem 5.4. *If* $(u^*, (y_1^*, y_2^*))$ *is an optimal pair for (DP2), and if* $p =$ (p_1, p_2) *is the solution to*

$$
\begin{cases}
\dfrac{\partial p_1}{\partial t} + d_1 \Delta p_1 = -r_1 p_1 + \mu_1 u^* y_2^* p_1 - \mu_2 u^*(x,t) y_2^*(x,t) \displaystyle\int_\Omega \ell(x',x) p_2(x',t) dx', \\[2.5ex]
\dfrac{\partial p_2}{\partial t} + d_2 \Delta p_2 = r_2 p_2 + \mu_1 u^* y_1^* p_1 - \mu_2 u^*(x,t) y_1^*(x,t) \displaystyle\int_\Omega \ell(x',x) p_2(x',t) dx', \\[1ex]
\hspace{5cm} (x,t) \in \Omega \times \in (0,T) \\[2ex]
\dfrac{\partial p_1}{\partial \nu}(x,t) = \dfrac{\partial p_2}{\partial \nu}(x,t) = 0, \hspace{2cm} (x,t) \in \partial\Omega \times (0,T) \\[2ex]
p_1(x,T) = p_2(x,T) = 1, \hspace{2.5cm} x \in \Omega,
\end{cases}
$$

then

$$
u^*(x,t) = \begin{cases}
0, \ if \ \mu_2 \displaystyle\int_\Omega \ell(x',x) p_2(x',t) dx' - \mu_1 p_1(x,t) < 0 \\[3ex]
1, \ if \ \mu_2 \displaystyle\int_\Omega \ell(x',x) p_2(x',t) dx' - \mu_1 p_1(x,t) > 0
\end{cases}
$$

a.e. on $\Omega \times (0,T)$.

It would be interesting to study the corresponding optimal control problems for all investigated models, with space structure this time (adding diffusion when necessary).

We now outline a Rosen-type projected gradient algorithm to approximate the optimal control of the above problem. Let $K = \{w \in L^2(\Omega \times (0,T)); \ 0 \le w(x,t) \le 1 \text{ a.e.}\}$.

A projected gradient algorithm for problem (DP2) (Rosen's method)

S0: Choose $u^{(0)} \in K$;
   Set j:=0;
S1: Compute $y^{(j)} = (y_1^{(j)}, y_2^{(j)})$ the solution to (5.20)–(5.23) with the input $u^{(j)}$:

$$\begin{cases} \dfrac{\partial y_1}{\partial t} - d_1 \Delta y_1 = r_1 y_1 - \mu_1 u^{(j)}(x,t) y_1 y_2, & x \in \Omega,\ t \in (0,T) \\[2mm] \dfrac{\partial y_2}{\partial t} - d_2 \Delta y_2 = -r_2 y_2 + \mu_2 \displaystyle\int_\Omega \ell(x,x') u^{(j)}(x',t) y_1(x',t) y_2(x',t) dx', \\[2mm] & x \in \Omega,\ t \in (0,T) \\[2mm] \dfrac{\partial y_1}{\partial \nu}(x,t) = \dfrac{\partial y_2}{\partial \nu}(x,t) = 0, & x \in \partial\Omega,\ t \in (0,T) \\[2mm] y_1(x,0) = y_{01}(x),\ \ y_2(x,0) = y_{02}(x), & x \in \Omega. \end{cases}$$

S2: Compute $p^{(j)} = (p_1^{(j)}, p_2^{(j)})$, the solution to

$$\begin{cases} \dfrac{\partial p_1}{\partial t} + d_1 \Delta p_1 = -r_1 p_1 + \mu_1 u^{(j)} y_2^{(j)} p_1 \\[2mm] \qquad - \mu_2 u^{(j)}(x,t) y_2^{(j)}(x,t) \displaystyle\int_\Omega \ell(x',x) p_2(x',t) dx', \ (x,t) \in \Omega \times \in (0,T) \\[2mm] \dfrac{\partial p_2}{\partial t} + d_2 \Delta p_2 = r_2 p_2 + \mu_1 u^{(j)} y_1^{(j)} p_1 \\[2mm] \qquad - \mu_2 u^{(j)}(x,t) y_1^{(j)}(x,t) \displaystyle\int_\Omega \ell(x',x) p_2(x',t) dx', \ (x,t) \in \Omega \times \in (0,T) \\[2mm] \dfrac{\partial p_1}{\partial \nu}(x,t) = \dfrac{\partial p_2}{\partial \nu}(x,t) = 0, & x \in \partial\Omega,\ t \in (0,T), \\[2mm] p_1(x,T) = p_2(x,T) = 1, & x \in \Omega. \end{cases}$$

S3: Compute $v^{(j)}$ according to the formula in Theorem 5.4:

$$v^{(j)}(x,t) = \begin{cases} 0 \text{ if } \mu_2 \displaystyle\int_\Omega \ell(x',x) p_2^{(j)}(x',t) dx' - \mu_1 p_1^{(j)}(x,t) < 0 \\[4mm] 1 \text{ if } \mu_2 \displaystyle\int_\Omega \ell(x',x) p_2^{(j)}(x',t) dx' - \mu_1 p_1^{(j)}(x,t) \geq 0 \end{cases}$$

a.e. on $\Omega \times (0,T)$.

S4: Compute $\lambda_j \in [0,1]$, the solution of the maximization problem

$$\max_{\lambda \in [0,1]} \Psi(\lambda u^{(j)} + (1-\lambda) v^{(j)}).$$

S5: Compute the new control $u^{(j+1)}$ by

$$u^{(j+1)} = \lambda_j u^{(j)} + (1 - \lambda_j) v^{(j)}.$$

S6: **(The stopping criterion)**
> if $\|u^{(j+1)} - u^{(j)}\| < \varepsilon$
>> then STOP ($u^{(j+1)}$ is the approximating control)
>> else $j := j + 1$; go to S1.

We recall that a convex combination of two bang-bang controls that take only the values 0 and 1, does not necessarily take only these two values. This is why it is better to replace the convex combination $\lambda u^{(j)} + (1 - \lambda)v^{(j)}$ from S4 and S5 by a convex combination of the switching points of $u^{(j)}$ and of $v^{(j)}$ as discussed in Section 4.3.

Bibliographical Notes and Remarks

There is an extensive literature devoted to the optimal control of diffusive models (see, e.g., [Lio72]). For models in population dynamics we refer to the monograph [Ani00]. For models regarding epidemics we refer to the monograph [Cap93] and references therein. For different models in economics as well as in physics and engineering, see [Bar94] and references therein. A variety of control problems for spatially structured epidemic systems can be found in [AC02], [AC09], [ACa09], and [ACv09].

For numerical methods for optimal control problems governed by PDEs we recommend [AN03].

Exercises

5.1. Derive the maximum principle for the following problem,

$$\text{Maximize } \int_0^T \int_0^A \int_\omega u(x, a, t) y^u(x, a, t) dx \, da \, dt,$$

where $u \in L^2(\Omega \times (0, A) \times (0, T))$, $0 \le u(x, a, t) \le L$ a.e., and y^u is the solution to

$$\begin{cases} \dfrac{\partial y}{\partial t} + \dfrac{\partial y}{\partial a} - \gamma \Delta y + \mu(a)y = -m(x)u(x,t)y(x,a,t), & (x, a, t) \in Q \\[2mm] \dfrac{\partial y}{\partial \nu}(x, a, t) = 0, & (x, t) \in \Sigma \\[2mm] y(x, 0, t) = \displaystyle\int_0^A \beta(a)y(x, a, t)da, & (x, t) \in \Omega \times (0, T) \\[2mm] y(x, a, 0) = y_0(x, a), & (x, t) \in \Omega \times (0, A), \end{cases}$$

where $Q = \Omega \times (0, A) \times (0, T)$, $\Sigma = \partial\Omega \times (0, A) \times (0, T)$. Here $\gamma > 0$, Ω, and ω satisfy the assumptions in Section 5.2, and β and μ satisfy the hypotheses in Section 4.1.

Hint. For details we refer to [Ani00].

5.2. Under the assumptions in Section 5.3, derive the maximum principle for the following problem:

$$\text{Maximize } \int_\Omega [y_1^u(x, T) + y_2^u(x, T)]dx, \qquad \textbf{(DP2}')$$

subject to $u \in L^2(\Omega \times (0, T))$, $0 \le u(x, t) \le 1$ a.e. $t \in (0, T)$, where (y_1^u, y_2^u) is the solution to

$$
\begin{cases}
\dfrac{\partial y_1}{\partial t} - d_1 \Delta y_1 = r_1 y_1 - \mu_1 u(x, t) y_1 y_2, & (x, t) \in \Omega \times \in (0, T) \\[2mm]
\dfrac{\partial y_2}{\partial t} - d_2 \Delta y_2 = -r_2 y_2 + \mu_2 u(x, t) y_1 y_2, & (x, t) \in \Omega \times \in (0, T) \\[2mm]
\dfrac{\partial y_1}{\partial \nu}(x, t) = \dfrac{\partial y_2}{\partial \nu}(x, t) = 0, & (x, t) \in \partial\Omega \times (0, T) \\[2mm]
y_1(x, 0) = y_{01}(x), \quad y_2(x) = y_{02}(x), & x \in \Omega.
\end{cases}
$$

Hint. (y_1^u, y_2^u) is the solution of the following initial-value problem (in $L^2(\Omega) \times L^2(\Omega)$).

$$
\begin{cases}
y'(t) = f(t, u(t), y(t)), & t \in (0, T) \\
y(0) = y_0,
\end{cases}
$$

where

$$
y_0 = \begin{pmatrix} y_{01} \\ y_{02} \end{pmatrix},
$$

$$
f(t, u, y) = Ay + \begin{pmatrix} r_1 y_1 - \mu_1 u y_1 y_2 \\ -r_2 y_2 + \mu_2 u y_1 y_2 \end{pmatrix}.
$$

Here

$$
y = \begin{pmatrix} y_1 \\ y_2 \end{pmatrix},
$$

and A is a linear unbounded operator. In fact A is defined by

$$
D(A) = \{w = (w_1, w_2) \in H^2(\Omega) \times H^2(\Omega); \ \frac{\partial w_1}{\partial \nu} = \frac{\partial w_2}{\partial \nu} = 0 \text{ on } \partial\Omega\},
$$

$$
Ay = \begin{pmatrix} d_1 \Delta y_1 \\ d_2 \Delta y_2, \end{pmatrix}, \quad y \in D(A).
$$

5.3. Under the assumptions in Section 5.3, derive the maximum principle for the following problem.

$$\text{Maximize} \int_{\Omega} [y_1^u(x,T) + y_2^u(x,T)]dx, \qquad \textbf{(DP2'')}$$

subject to $u \in L^2(\Omega \times (0,T))$, $0 \le u(x,t) \le 1$ a.e. $t \in (0,T)$, where (y_1^u, y_2^u) is the solution to

$$\begin{cases} \dfrac{\partial y_1}{\partial t} - d_1 \Delta y_1 = r_1 y_1 \left(1 - \dfrac{y_1}{k}\right) - \mu_1 u(x,t) y_1 y_2, & (x,t) \in \Omega \times (0,T) \\[2mm] \dfrac{\partial y_2}{\partial t} - d_2 \Delta y_2 = -r_2 y_2 + \mu_2 u(x,t) y_1 y_2, & (x,t) \in \Omega \times (0,T) \\[2mm] \dfrac{\partial y_1}{\partial \nu}(x,t) = \dfrac{\partial y_2}{\partial \nu}(x,t) = 0, & (x,t) \in \partial \Omega \times (0,T) \\[2mm] y_1(x,0) = y_{01}(x), \quad y_2(x) = y_{02}(x), & x \in \Omega. \end{cases}$$

Here k is a positive constant.

A

Appendices

A.1 Elements of functional analysis

A.1.1 The Lebesgue integral

Let the real numbers be $a_i < b_i$, $i = 1, 2, \ldots, m$. A set of the form

$$C = \prod_{i=1}^{m} [a_i, b_i]$$

is called a *cell*.

Denote by $\mathcal{F}$ the set of all finite unions of disjoint cells. If $F = \bigcup_{i=1}^{n} C_i \in \mathcal{F}$, where C_i are disjoint cells, then we define the *measure* of F by

$$\mu(F) = \sum_{i=1}^{n} \mu(C_i),$$

where $\mu(C_i)$ is the volume of C_i. Let $\Omega \subset \mathbb{R}^m$ be an open subset. We define the measure of Ω by

$$\mu(\Omega) = \sup\{\mu(F); \ F \subset \Omega, \ F \in \mathcal{F}\}. \tag{A.1}$$

Let $K \subset \mathbb{R}^m$ be a compact subset. We define the measure of K by

$$\mu(K) = \inf\{\mu(\Omega); \ K \subset \Omega, \ \Omega \text{ open}\}. \tag{A.2}$$

Let now $A \subset \mathbb{R}^m$ be a bounded set. We define the *outer measure* of A by

$$\mu^*(A) = \inf\{\mu(\Omega); \ A \subset \Omega, \ \Omega \text{ open}\}$$

and the *inner measure* of A by

$$\mu_*(A) = \sup\{\mu(K); \ K \subset A, \ K \text{ compact}\}.$$

By definition the bounded set A is *measurable* if $\mu^*(A) = \mu_*(A)$. The common value is denoted by $\mu(A)$ or $meas(A)$ and it is called the *measure* of A.

Now let $A \subset \mathbb{R}^m$ be an unbounded set. We introduce the ball

$$B_r = \{x \in \mathbb{R}^m; \ \|x\| < r\},$$

where $\|\cdot\|$ denotes the Euclidean norm in $\mathbb{R}^m$.

By definition the unbounded set A is *measurable* if the bounded set $A \cap B_r$ is measurable for any $r > 0$.

We say that a certain property holds almost everywhere on A (a.e. on A) if the property holds on $A \backslash A_0$, where $A_0 \subset A$ is measurable and $\mu(A_0) = 0$.

Now let $f : \mathbb{R}^m \to [-\infty, +\infty]$. The function f is said to be *measurable* if for each $\lambda \in \mathbb{R}$, the level set $\{x \in \mathbb{R}^m; \ f(x) \le \lambda\}$ is measurable.

Let $f : \mathbb{R}^m \to \mathbb{R}$. The function f is *finitely valued* if there exist $a_i \in \mathbb{R}$, $i = 1, 2, \ldots, n$, and $A_i \subset \mathbb{R}^m$, $i = 1, 2, \ldots, n$, measurable and mutually disjoint subsets such that

$$f(x) = \sum_{i=1}^{n} a_i \chi_{A_i}(x), \quad x \in \mathbb{R}^m, \tag{A.3}$$

where χ_{A_i} is the characteristic function of A_i.

If $f : \mathbb{R}^m \to [-\infty, +\infty]$ is measurable then there is a sequence of finitely valued functions that is a.e. convergent to f on $\mathbb{R}^m$. The integral $I(f)$ of the finitely valued function f defined by formula (A.3) is, by definition,

$$I(f) = \sum_{i=1}^{n} a_i \mu(A_i). \tag{A.4}$$

Let $f : \mathbb{R}^m \to [0, +\infty]$. The function f is *Lebesgue integrable* if

$$I(f) = \sup \{I(\varphi); \ 0 \le \varphi \le f, \ \varphi \text{ finitely valued}\} < +\infty. \tag{A.5}$$

Then $I(f)$ is called the *Lebesgue integral* of f. It is also denoted $\int f(x)dx$.

If $A \subset \mathbb{R}^m$ is a measurable subset, then the integral of f on A is defined by

$$\int_A f(x)dx = I(\chi_A f),$$

where χ_A is the characteristic function of A.

Now let $f : \mathbb{R}^m \to [-\infty, +\infty]$. We define the *positive part* of f to be $f^+ : \mathbb{R}^m \to [0, +\infty]$ by $f^+(x) = \max \{f(x), 0\}$ and the *negative part* of f to be $f^- : \mathbb{R}^m \to [0, +\infty]$ by $f^-(x) = \max\{-f(x), 0\}$. A simple calculus shows that $f = f^+ - f^-$, and $|f| = f^+ + f^-$.

The function f is *Lebesgue integrable* if f^+ and f^- are both Lebesgue integrable. Then the integral of f is defined by

$$\int f(x)dx = \int f^+(x)dx - \int f^-(x)dx.$$

A first result refers to sequences of measurable functions.

Theorem A.1. *If* $\{f_k\}$ *is a sequence of measurable functions, and*

$$f(x) = \lim_{k \to +\infty} f_k(x) \quad a.e. \ x \in \mathbb{R}^m,$$

then f *is measurable.*

We also have the following theorem.

Theorem A.2. *Let* $f : \mathbb{R}^m \to [-\infty, +\infty]$ *be a measurable function. If* f *is integrable, then*

$$\left| \int f(x)dx \right| \leq \int |f(x)|dx.$$

The following result is known as the *Lebesgue dominated convergence theorem.*

Theorem A.3. *Let* $\{f_k\}$ *be a sequence of measurable functions and* g *an integrable function such that*

$$|f_k(x)| \leq g(x) \quad a.e. \ x \in \mathbb{R}^m,$$

and

$$f(x) = \lim_{k \to +\infty} f_k(x) \quad a.e. \ x \in \mathbb{R}^m.$$

Then f *is integrable and*

$$\int f(x)dx = \lim_{k \to +\infty} \int f_k(x)dx.$$

A.1.2 L^p spaces

Let $\Omega \subset \mathbb{R}^m$ be a bounded measurable subset and $1 \leq p < +\infty$. Then $L^p(\Omega)$ is the space of all measurable functions (classes of functions) $f : \Omega \to \mathbb{R}$ such that $|f|^p$ is Lebesgue integrable on Ω. $L^p(\Omega)$ is a *Banach space* with the norm

$$\|f\|_{L^p(\Omega)} = \left(\int_\Omega |f(x)|^p dx \right)^{1/p}. \tag{A.6}$$

Moreover, for $1 < p < +\infty$ the dual of the Banach space $L^p(\Omega)$ is $L^q(\Omega)$, where $1/p + 1/q = 1$, and $L^p(\Omega)$ is reflexive (for details see [Bre83]).

For $p = 2$ the space $L^2(\Omega)$ is a Hilbert space endowed with the scalar product (inner product)

$$(f, g)_{L^2(\Omega)} = \int_\Omega f(x)g(x)dx. \qquad (A.7)$$

For $p = 1$ we have the particular case of $L^1(\Omega)$, the space of all Lebesgue integrable functions on Ω. The corresponding norm is

$$\|f\|_{L^1(\Omega)} = \int_\Omega |f(x)|dx.$$

For $p = +\infty$ we have $L^\infty(\Omega)$, the space of essentially bounded functions on Ω.

$$\|f\|_{L^\infty(\Omega)} = \text{Ess sup}\{|f(x)|; \ x \in \Omega\}.$$

Recall that f is essentially bounded if there exists a constant α such that $|f(x)| \leq \alpha$ a.e. on Ω. The infimum of such constants α is denoted by $\text{Ess sup}\{|f(x)|; \ x \in \Omega\}$.

We also recall the *Hölder inequality*. Let $f \in L^p(\Omega)$ and $g \in L^q(\Omega)$, where $p, q \in (1, +\infty)$, $1/p + 1/q = 1$. Then

$$\int_\Omega |f(x)g(x)|dx \leq \left(\int_\Omega |f(x)|^p dx\right)^{1/p} \cdot \left(\int_\Omega |g(x)|^q dx\right)^{1/q}. \qquad (A.8)$$

Moreover for $f_i \in L^{p_i}(\Omega)$, $i = 1, 2, \ldots, n$, with $\sum_{i=1}^n 1/p_i = 1$, $(p_i > 1)$, we have

$$\left|\int_\Omega \prod_{i=1}^n f_i(x)dx\right| \leq \prod_{i=1}^n \left(\int_\Omega |f_i(x)|^{p_i} dx\right)^{1/p_i}.$$

We consider now the extension of the integral to a Banach space-valued function defined on a real interval. To begin let $f : \mathbb{R} \to V$, where V is a (real) Banach space. We denote by $\|\cdot\|$ the norm of V. Let $\{A_1, A_2, \ldots, A_n\}$ be a finite set of measurable and mutually disjoint subsets of $\mathbb{R}$ each having finite measure and $\{a_1, a_2, \ldots, a_n\}$ a corresponding set of elements of V. We say that f given by

$$f(t) = \sum_{i=1}^n a_i \chi_{A_i}(t), \quad t \in \mathbb{R},$$

where χ_{A_i} is the characteristic function of A_i, is a finitely valued function.

Now let $a < b$ and $f : [a, b] \to V$. The function f is *strongly measurable* on $[a, b]$ if there exists a sequence $\{f_k\}$ of finitely valued functions, $f_k : [a, b] \to V$, such that

$$\lim_{k \to +\infty} \|f_k(t) - f(t)\| = 0 \ \text{ a.e. } t \in [a, b]. \qquad (A.9)$$

We say that f is *Bochner integrable* on $[a, b]$ if there exists a sequence $\{f_k\}$ of finitely valued functions, $f_k : [a, b] \to V$, such that (A.9) is satisfied (f is strongly measurable) and

$$\lim_{k \to +\infty} \int_a^b \|f_k(t) - f(t)\| dt = 0.$$

A necessary and sufficient condition that $f : [a, b] \to V$ is Bochner integrable is that f is strongly measurable, and $\int_a^b \|f(t)\| dt < +\infty$.

Let $A \subset \mathbb{R}$ be a measurable set. Then

$$\left\| \int_A f(t) dt \right\| \leq \int_A \|f(t)\| dt.$$

We denote by $L^p(a, b; V)$ the linear space of (classes of) measurable functions $f : [a, b] \to V$ such that $\|f(\cdot)\| \in L^p(a, b)$; that is,

$$\int_a^b \|f(t)\|^p dt < +\infty \text{ for } 1 \leq p < +\infty,$$

and

$$\text{Ess sup } \{\|f(t)\|;\ t \in [a, b]\} < +\infty \text{ for } p = +\infty.$$

The corresponding norm is, respectively,

$$\|f\| = \left(\int_a^b \|f(t)\|^p dt \right)^{1/p} \text{ for } f \in L^p(a, b; V), \quad 1 \leq p < +\infty, \qquad (A.10)$$

and

$$\|f\| = \text{Ess sup}\{\|f(t)\|;\ t \in [a, b]\} \text{ for } f \in L^\infty(a, b; V). \qquad (A.11)$$

If the Banach space V is reflexive and $1 < p < +\infty$, then the dual space of $L^p(a, b; V)$ is $L^q(a, b; V^*)$ where $1/p + 1/q = 1$ and V^* is the dual space of V.

For PDEs usually we consider $L^p(0, T; V)$ ($p \in [1, +\infty]$); that is, the time interval $[a, b]$ becomes $[0, T]$. A particular useful case is $V = L^r(\Omega)$ with $\Omega \subset \mathbb{R}^m$ a nonempty bounded open set. Another one is $V = \mathbb{R}^N$. The set $L^p_{loc}(a, b; V)$ is defined by

$$L^p_{loc}(a, b; V) = \{f \in L^p(c, d; V);\ \text{for any } c,\ d \text{ such that } a < c < d < b\}.$$

For $p = 1$ we say that f is locally integrable.

For any $A, T \in (0, +\infty)$, we finally define

$$L^\infty_{loc}([0, A)) = \{f : [0, A) \to \mathbb{R};\ f \in L^\infty(0, \tilde{A}) \text{ for any } \tilde{A} \in (0, A)\}$$

(the set of all locally essentially bounded functions from $[0, A)$ to $\mathbb{R}$), and

$$L^\infty_{loc}([0, A) \times [0, T]) = \{f : [0, A) \times [0, T] \to \mathbb{R};\ f \in L^\infty((0, \tilde{A}) \times (0, T))$$
$$\text{for any } \tilde{A} \in (0, A)\}$$

(the set of all locally essentially bounded functions from $[0, A) \times [0, T]$ to $\mathbb{R}$).

A.1.3 The weak convergence

Let H be a real Hilbert space endowed with the scalar product (inner product) $(\cdot, \cdot)$ and the corresponding norm $\| \cdot \|$ defined by $\|x\| = (x, x)^{1/2}$. We assume that H is identified with its own dual (by Riesz's theorem).

We say that the sequence $\{x_n\} \subset H$ is *weakly convergent* to $x \in H$ if

$$\lim_{n \to +\infty} (x_n, y) = (x, y) \text{ for any } y \in H.$$

The following results are valid.

Theorem A.4. *Let $\{x_n\}$ be a sequence in the Hilbert space H and $x \in H$.*

(1) If $\{x_n\}$ is strongly convergent to x, that is, $\lim_{n \to +\infty} \|x_n - x\| = 0$, then $\{x_n\}$ is weakly convergent to x.
(2) If $\{x_n\}$ is weakly convergent to x, then the sequence $\|x_n\|$ is bounded and $\|x\| \leq \lim \inf \|x_n\|$.
(3) If $\{x_n\}$ is weakly convergent to x and $\{y_n\} \subset H$ is strongly convergent to $y \in H$, then $\{(x_n, y_n)\}$ is convergent to (x, y) in $\mathbb{R}$.

For a finite-dimensional space (for instance, $\mathbb{R}^m$) the weak and strong convergence are equivalent to the usual convergence.

We say that a subset $\mathcal{A} \subset H$ is *sequentially weakly compact* if every sequence $\{x_n\} \in \mathcal{A}$ contains a weakly convergent subsequence. For finite-dimensional spaces the sequentially weakly compact sets are the relatively compact sets and the weak convergence is the usual one. A very important property of Hilbert spaces, and more generally of reflexive Banach spaces, is that every bounded subset is sequentially weakly compact. Namely, the following result is valid.

Theorem A.5. *Let $\{x_n\} \subset H$ be a bounded sequence. Then there exists a subsequence $\{x_{n_k}\} \subset \{x_n\}$ that is weakly convergent to an element of H.*

If we apply Theorem A.5 to the Hilbert space $H = L^2(\Omega)$ (where $\Omega \subset \mathbb{R}^m$, $m \in \mathbb{N}^*$, is a nonempty bounded open subset) it follows that every bounded subset of $L^2(\Omega)$ is sequentially weakly compact. If $\{f_n\} \subset L^2(\Omega)$ is a bounded sequence of functions, that is, there exists $M > 0$ such that

$$\|f_n\|_{L^2(\Omega)} \leq M \text{ for any } n,$$

then there exists the subsequence $\{f_{n_k}\} \subset \{f_n\}$ and $f \in L^2(\Omega)$ such that

$$\lim_{k \to +\infty} \int_\Omega f_{n_k}(x)\varphi(x)dx = \int_\Omega f(x)\varphi(x)dx \text{ for any } \varphi \in L^2(\Omega).$$

A.1.4 The normal cone

Let K be a closed convex subset of the real Hilbert space V endowed with the inner product $(\cdot, \cdot)$ and $u \in K$. The set

$$N_K(u) = \{w \in V; \ (v - u, w) \le 0 \text{ for any } v \in K\}$$

is called the *normal cone* to K at u.

Example A.6. If $V = \mathbb{R}$, $K = [a, b]$ $(a, b \in \mathbb{R}, \ a < b)$, then

$$N_K(u) = \begin{cases} \mathbb{R}^+, & u = b \\ \{0\}, & a < u < b \\ \mathbb{R}^-, & u = a. \end{cases}$$

Example A.7. Let $\Omega \subset \mathbb{R}^m$ $(m \in \mathbb{N}^*)$ be an open and bounded subset, and $\zeta_1, \zeta_2 \in L^2(\Omega)$, with $\zeta_1(x) < \zeta_2(x)$ a.e. $x \in \Omega$. Define

$$K = \{y \in L^2(\Omega); \ \zeta_1(x) \le y(x) \le \zeta_2(x) \text{ a.e. } x \in \Omega\}.$$

K is a closed convex subset of $L^2(\Omega)$. Let $y \in K$. Then for any $z \in N_K(y)$ we have:

- $z(x) \le 0$ for $y(x) = \zeta_1(x)$;
- $z(x) \ge 0$ for $y(x) = \zeta_2(x)$;
- $z(x) = 0$ for $\zeta_1(x) < y(x) < \zeta_2(x)$,

a.e. in Ω.

Indeed, consider $A = \{x \in \Omega; \ y(x) = \zeta_1(x)\}$. Let us prove that $z(x) \le 0$ a.e. $x \in A$. Assume by contradiction that there exists a subset $\tilde{A} \subset A$ such that $\mu(\tilde{A}) > 0$ and $z(x) > 0$ a.e. $x \in \tilde{A}$.
Because $z \in N_K(y)$, we get that

$$\int_\Omega z(x)(h(x) - y(x))dx \le 0,$$

for any $h \in K$.
We may choose $h \in K$ such that

$$h(x) = \zeta_1(x) \text{ a.e. } x \in \Omega \setminus \tilde{A},$$

$$h(x) > \zeta_1(x) \text{ a.e. } x \in \tilde{A}.$$

Thus

$$\int_\Omega z(x)(h(x) - y(x))dx = \int_{\tilde{A}} z(x)(h(x) - y(x))dx > 0,$$

which is absurd. In conclusion, $z(x) \le 0$ a.e. $x \in A$.
In the same manner it follows that $z(x) \ge 0$ for almost any $x \in \Omega$ such that $y(x) = \zeta_2(x)$, and $z(x) = 0$ for almost any $x \in \Omega$ such that $\zeta_1(x) < y(x) < \zeta_2(x)$.

A.1.5 The Gâteaux derivative

Consider the real normed spaces V and H and the operator $F : D \subseteq V \to H$. We define the *directional derivative* of F at $x \in \text{int}(D)$ along the direction $h \in V$ by

$$F'(x, h) = \lim_{\lambda \to 0} \frac{F(x + \lambda h) - F(x)}{\lambda},$$

if the limit exists.

It is easy to see that $F'(x, 0) = 0$ for any $x \in \text{int}(D)$ and that the operator $h \to F'(x, h)$ is homogeneous:

$$F'(x, \alpha h) = \alpha F'(x, h) \text{ for any } \alpha \in \mathbb{R}.$$

However, the operator above does not have the additivity property and therefore it is not necessarily linear.

Let us introduce now the left and right directional derivatives. The right directional derivative of F at $x \in \text{int}(D)$ along the direction $h \in V$ is given by

$$F'_+(x, h) = \lim_{\lambda \to 0^+} \frac{F(x + \lambda h) - F(x)}{\lambda}$$

if the limit exists. The left directional derivative of F at $x \in \text{int}(D)$ along the direction $h \in V$ is given by

$$F'_-(x, h) = \lim_{\lambda \to 0^-} \frac{F(x + \lambda h) - F(x)}{\lambda}$$

if the limit exists. It is easy to see that $F'_-(x, h) = -F'_+(x, -h)$.

Proposition A.8. *F has the directional derivative at $x \in \text{int}(D)$ along the direction h if and only if both the right and left directional derivatives at x along the direction h exist and are equal. In such a case we have*

$$F'(x, h) = F'_+(x, h) = F'_-(x, h).$$

We can now state the definition of the Gâteaux derivative.

If for $x \in \text{int}(D)$ the derivative $F'(x, h)$ exists along any direction h and if the operator $h \mapsto F'(x, h)$ is linear and continuous, then we say that F is Gâteaux differentiable (weakly differentiable) at x. In such a case $F'(x) \in L(V, H)$ defined by

$$F'(x)h = \lim_{\lambda \to 0} \frac{F(x + \lambda h) - F(x)}{\lambda},$$

is called the Gâteaux derivative (the weak derivative) of F at x. It is also denoted by $F_x(x)$ or by $dF(x)$.

An equivalent definition is the following one. F is weakly differentiable at $x \in \text{int}(D)$ if there exists the linear and continuous operator $A : V \to H$ such that

$$\lim_{\lambda \to 0} \frac{\|F(x + \lambda h) - F(x) - \lambda A h\|}{\lambda} = 0$$

for any $h \in V$. In such a case $F'(x) = A$ and

$$\lim_{\lambda \to 0} \frac{F(x + \lambda h) - F(x)}{\lambda} = Ah. \tag{A.12}$$

We also have the following.

Lemma A.9. *The linear operator A defined above is unique.*

In the finite-dimensional case, $F : D \subseteq \mathbb{R}^n \to \mathbb{R}^m$, we make some computational remarks. Consider that $F = (f_1, f_2, \ldots, f_m)^T$, where $f_i : D \subset \mathbb{R}^n \to \mathbb{R}$, $i = 1, 2, \ldots, m$. We take $h = e_j$ in (A.12), where e_j is the normal unit vector, and because A is a real matrix with m rows and n columns we get component by component

$$\lim_{\lambda \to 0} \frac{f_i(x + \lambda e_j) - f_i(x)}{\lambda} = a_{ij},$$

and hence $a_{ij} = \partial f_i / \partial x_j(x)$, $j = 1, 2, \ldots, n$, $i = 1, 2, \ldots, m$. We may conclude that the matrix representation of the operator $A = F'(x)$ is the Jacobi matrix

$$F'(x) = [a_{ij}]_{m \times n}, \ a_{ij} = \frac{\partial f_i}{\partial x_j}(x).$$

It is also denoted by $\dfrac{\partial F}{\partial x}(x)$ or $F_x(x)$.

Consider the particular case in which $m = 1$. Then $f : D \subseteq \mathbb{R}^n \to \mathbb{R}$ and

$$f'(x) = \left(\frac{\partial f}{\partial x_i}(x) \right)_{i=1,\ldots,n}$$

is the gradient of f at x. However, $f'x$ is usually considered as a column vector, and it is also denoted by $\dfrac{\partial f}{\partial x}$, $f_x(x)$, $\nabla f(x)$, $\nabla_x f(x)$, or $\mathrm{grad} f(x)$.

Let us recall that

- The existence of the Jacobi matrix, that is, the existence of all partial derivatives, at $x \in D$ does not imply the weak differentiability at x.
- The existence of the weak derivative at $x \in D$ does not imply the continuity at x.

For more elements of functional analysis we recommend [Ada75], [Bre83], and [BP86].

A.2 Bellman's lemma

We present here two useful results related to integral inequalities. We begin with Gronwall's lemma.

Lemma A.10. *Let* $x : [a, b] \to \mathbb{R}$ *(a, b* $\in \mathbb{R}$, *a < b) be a continuous function,* $\varphi \in L^\infty(a, b)$, *and* $\psi \in L^1(a, b)$, $\psi(t) \geq 0$ *a.e.* $t \in (a, b)$. *If*

$$x(t) \leq \varphi(t) + \int_a^t \psi(s)x(s)ds,$$

for almost any $t \in [a, b]$, *then*

$$x(t) \leq \varphi(t) + \int_a^t \varphi(s)\psi(s) \exp\left(\int_s^t \psi(\tau)d\tau \right) ds,$$

for almost any $t \in [a, b]$.

Proof. Denote

$$y(t) = \int_a^t \psi(s)x(s)ds, \quad t \in [a, b].$$

The equality

$$y'(t) = \psi(t)x(t) \text{ a.e. } t \in (a, b),$$

and the hypotheses imply that

$$y'(t) \leq \psi(t)\varphi(t) + \psi(t)y(t) \text{ a.e. } t \in (a, b).$$

We multiply the last inequality by $\exp\left(- \int_a^t \psi(s)ds \right)$, and obtain that

$$\frac{d}{dt}\left[y(t)\exp\left(- \int_a^t \psi(s)ds \right) \right] \leq \psi(t)\varphi(t)\exp\left(- \int_a^t \psi(s)ds \right)$$

a.e. $t \in (a, b)$. By integration we infer that

$$y(t) \leq \int_a^t \psi(s)\varphi(s)\exp\left(\int_s^t \psi(\tau)d\tau \right) ds,$$

for any $t \in [a, b]$, and consequently we obtain the conclusion of the lemma.

If in addition, $\varphi(t) = M$ for each $t \in [a, b]$ (φ is a constant function), by Lemma A.10 we may deduce the following Bellman's lemma.

Lemma A.11. *If* $x \in C([a, b])$, $\psi \in L^1(a, b)$, $\psi(t) \geq 0$ *a.e.* $t \in (a, b)$, $M \in \mathbb{R}$, *and*

$$x(t) \leq M + \int_a^t \psi(s)x(s)ds,$$

for each $t \in [a, b]$, *then*

$$x(t) \leq M \exp\left(\int_a^t \psi(s)ds \right),$$

for each $t \in [a, b]$.

A.3 Existence and uniqueness of Carathéodory solution

Let X be a real Banach space with norm $\|\cdot\|$. Let $a, b \in \mathbb{R}$, $a < b$. We say that function $x : [a, b] \to X$ is *continuous* at $t_0 \in [a, b]$ if the real function $g : [a, b] \to \mathbb{R}$,

$$g(t) = \|x(t) - x(t_0)\|, \quad t \in [a, b]$$

is continuous at t_0. If x is continuous at each $t \in [a, b]$, we say that x is continuous on $[a, b]$.

Denote by $C([a, b]; X)$ the space of all continuous functions $x : [a, b] \to X$. This is a Banach space with the norm

$$\|x\|_{C([a,b];X)} = \max\{\|x(t)\|; \ t \in [a, b]\}.$$

We say that the function $x : [a, b] \to X$ is *absolutely continuous* on $[a, b]$ if for each $\varepsilon > 0$, there exists $\delta(\varepsilon) > 0$ such that

$$\sum_{k=1}^{N} \|x(t_k) - x(s_k)\| < \varepsilon,$$

whenever $\sum_{k=1}^{N} |t_k - s_k| < \delta(\varepsilon)$ and $(t_k, s_k) \cap (t_j, s_j) = \emptyset$, for $k \neq j$ $(t_k, s_k \in [a, b]$, for any $k \in \{1, 2, \ldots, N\})$.

We denote by $AC([a, b] : X)$ the space of all absolutely continuous functions $x : [a, b] \to X$.

We say that the function $x : [a, b] \to X$ is *differentiable* at $t_0 \in [a, b]$ if there exists $x'(t_0) \in X$, the derivative of x at t_0, such that

$$\lim_{t \to t_0} \left\| \frac{1}{t - t_0}(x(t) - x(t_0)) - x'(t_0) \right\| = 0.$$

The derivative x' is also denoted by dx/dt. We say that x is differentiable on $[a, b]$ if x is differentiable at each $t \in [a, b]$.

We denote by $C^k([a, b]; X)$ $(k \in \mathbb{N}^*)$ the space of all functions $x : [a, b] \to X$, k times differentiable, with continuous kth derivative.

Theorem A.12. *Let X be a reflexive Banach space. If $x \in AC([a, b]; X)$, then x is almost everywhere differentiable on $[a, b]$, and*

$$x(t) = x(a) + \int_a^t x'(s)ds \ \text{for any } t \in [a, b].$$

For the proof see [Bar93].

Theorem A.13. *If* $x \in C([a,b]; X)$ *and* $x' \in L^1(a,b; X)$, *then* $x \in AC([a,b]; X)$.

For the proof see [Bar98].

Let us consider the following Cauchy problem (IVP):

$$\begin{cases} x'(t) = f(t, u(t), x(t)), & t \in (0, T) \\ x(0) = x_0, \end{cases} \tag{A.13}$$

where $T > 0$, $x_0 \in \mathbb{R}^N$ ($N \in \mathbb{N}^*$), and $u \in L^1(0, T; \mathbb{R}^m)$ ($m \in \mathbb{N}^*$). Here $f : [0, T] \times \mathbb{R}^m \times \mathbb{R}^N \to \mathbb{R}^N$, is measurable in t and continuous in (u, x) and satisfies

$$\|f(t, u, x)\| \leq a_1(t)\|x\| + a_2(t),$$
$$\|f(t, u, x) - f(t, u, y)\| \leq b(t)\|x - y\|,$$

a.e. $(t, u, x), (t, u, y) \in (0, T) \times \mathbb{R}^m \times \mathbb{R}^N$, where $a_1, a_2, b \in L^1(0, T)$ and $\|\cdot\|$ denotes the norm of $\mathbb{R}^N$.

By a Carathéodory solution (or simply solution) to Problem (A.13) we mean an absolutely continuous function $x : [0, T] \to \mathbb{R}^N$ such that

$$\begin{cases} x'(t) = f(t, u(t), x(t)) & \text{a.e. } t \in (0, T) \\ x(0) = x_0. \end{cases} \tag{A.14}$$

Note that $\mathbb{R}^N$ is a reflexive Banach space so that Theorem A.12 applies.

Theorem A.14. *Problem (A.13) admits a unique solution.*

For the proof see [CL55].

A.4 Runge–Kutta methods

Let us consider the Initial-Value Problem (IVP) (1.4) from Chapter 1, that is,

$$\begin{cases} y'(x) = f(x, y(x)), \\ y(x_0) = y_0, \end{cases} \tag{A.15}$$

together with the related hypotheses (see Theorem 1.1 in Chapter 1). Let $[\alpha, \beta]$ be a real interval, $\varphi \in C([\alpha, \beta])$ some given function, and the general numerical integration formula

$$\int_\alpha^\beta \varphi(x)dx = \sum_{i=1}^q c_i\varphi(z_i) + R_q(\varphi). \tag{A.16}$$

Here $c_i \in \mathbb{R}$, $i = 1, 2, \ldots, q$, are the coefficients of the formula, $z_i \in [\alpha, \beta]$, $i = 1, 2, \ldots, q$, are the knots, and $R_q(\varphi)$ is the remainder. To use the numerical integration formula (A.16) means to neglect the remainder and to get the approximation

$$\int_\alpha^\beta \varphi(x)dx \approx \sum_{i=1}^q c_i \varphi(z_i).$$ (A.17)

Coming back to the IVP (A.15), we introduce the following grid $x_0 < x_1 < \cdots < x_N$ of equidistant knots of step h and we now integrate the corresponding ODE successively on the intervals $[x_0, x_1]$, $[x_1, x_2]$, ..., $[x_{N-1}, x_N]$. Let us now integrate it on some interval $[x_n, x_{n+1}]$. We get

$$y(x_{n+1}) = y(x_n) + \int_{x_n}^{x_{n+1}} f(x, y(x))dx.$$ (A.18)

We denote in the sequel by y_j the numerical approximation of $y(x_j)$, where y is the solution of problem (A.15), for any j. Formula (A.18) raises the following problem. We have integrated (A.15) up to x_n and we want to advance now from x_n to x_{n+1}, that is, to integrate the IVP

$$\begin{cases} y'(x) = f(x, y(x)), & x \in (x_n, x_{n+1}] \\ y(x_n) = y_n, \end{cases}$$

where y_n is the approximation already known for $y(x_n)$. However, we have no numerical information about the solution $y(x)$ on the interval $(x_n, x_{n+1}]$ to apply a numerical integration formula. Hence we first introduce the unknown knots of the numerical integration formula

$$x_{n,i} = x_n + \alpha_i h \in [x_n, x_{n+1}],$$

for $i = 1, 2, \ldots, q$. We assume of course that $x_n = x_{n,1} \le x_{n,2} \le \cdots \le x_{n,q} = x_{n+1}$. By using a formula like (A.17) for (A.18) we get that

$$y_{n+1} = y_n + h \sum_{i=1}^q p_i f(x_{n,i}, y_{n,i}).$$ (A.19)

Comparing with formula (A.17) we see that $x_{n,i}$ stands for z_i, $c_i = hp_i$, and $y_{n,i}$ denotes the corresponding approximation of $y(x_{n,i})$. We now integrate the ODE from x_n to $x_{n,i}$ and obtain

$$y(x_{n,i}) = y(x_n) + \int_{x_n}^{x_{n,i}} f(x, y(x))dx.$$ (A.20)

We use the knots $x_{n,j}$, $j = 0, 1, \ldots, i-1$, on the interval $[x_n, x_{n,i}]$ and we get the following numerical integration formula for (A.20):

$$y_{n,i} = y_n + h \sum_{j=1}^{i-1} b_{ij} f(x_{n,j}, y_{n,j}).$$ (A.21)

Because $x_{n,1} = x_n$, we also have $y_{n,1} = y_n$.

The problem now is to find the unknowns p_i, α_i, and b_{ij}. Let us first denote

$$k_i = k_i(h) = hf(x_{n,i}, y_{n,i}), \quad i = 1, \ldots, q. \tag{A.22}$$

Therefore formula (A.19) may be rewritten as

$$y_{n+1} = y_n + \sum_{i=1}^{q} p_i k_i, \tag{A.23}$$

and (A.21) reads

$$y_{n,i} = y_n + \sum_{j=1}^{i-1} b_{ij} k_j, \quad i = 1, \ldots, q. \tag{A.24}$$

Let us assume that $f \in C^m(D)$ $(m \in \mathbb{N}^*)$, where (we recall from Section 1.4):

$$D = \{(x, y) \in \mathbb{R}^2; \ |x - x_0| \le a, \ |y - y_0| \le b\} \quad (a, \ b > 0).$$

Then, according to Remark 1.3 from Section 1.4, the solution of Problem (A.15) satisfies $y \in C^{m+1}([x_0, x_0 + \delta])$. We take, of course, $x_N \le x_0 + \delta$. By Taylor's formula one has

$$y(x_{n+1}) = y(x_n + h) = y(x_n) + \sum_{j=1}^{m} \frac{h^j}{j!} y^{(j)}(x_n) + \frac{h^{m+1}}{(m+1)!} y^{(m+1)}(x_n + \theta h), \tag{A.25}$$

where $0 < \theta < 1$. On the other hand y_{n+1} from Formula (A.23) depends on h by the contributions of the functions $k_i(h)$. We therefore have the Taylor expansion

$$y_{n+1}(h) = y_{n+1}(0) + \sum_{j=1}^{m} \frac{h^j}{j!} y_{n+1}^{(j)}(0) + \frac{h^{m+1}}{(m+1)!} y_{n+1}^{(m+1)}(\xi), \tag{A.26}$$

where $0 < \xi < h$.

The idea behind Runge–Kutta (RK) methods is the following. Find the unknown coefficients $(p_i, \alpha_i,$ and $b_{ij})$ such that the expansions (A.25) and (A.26) have equal terms as much as possible; that is, the numerical approximation (A.26) of (A.25) is as good as possible. It is quite clear that the number of coefficients in formulae (A.23) and (A.24), the coefficients to be determined, depends on q which is called the *order of the RK method*. We demonstrate that the first term in both expansions (A.25) and (A.26) is the same. From the initial condition of the IVP integrated on the interval $[x_n, x_{n+1}]$ we have $y(x_n) = y_n$. By Formula (A.22) we get $k_i(0) = 0$, and Formula (A.23) leads to $y_{n+1}(0) = y_n$. Hence $y(x_n) = y_{n+1}(0)$. Then we introduce the following conditions:

$$y^{(j)}(x_n) = y_{n+1}^{(j)}(0), \quad j = 1, \ldots, s. \tag{A.27}$$

By taking into account (A.27), the truncation error becomes

$$y(x_{n+1}) - y_{n+1}(h) = \frac{h^{s+1}}{(s+1)!}\left[y^{(s+1)}(x_n) - y_{n+1}^{(s+1)}(0)\right] + O(h^{s+2}). \quad \text{(A.28)}$$

Solving the system (A.27) we get the values of the unknowns p_i, α_i, and b_{ij} such that formulae (A.23) and (A.24) lead to the approximation of $y(x_{n+1})$ with the precision given by formula (A.28).

Deriving Runge–Kutta methods. As already asserted we always consider $x_{n,1} = x_n$ and $y_{n,1} = y_n$. Therefore $\alpha_1 = 0$ and $b_{11} = 0$.

Runge–Kutta methods of order 1 $(q = 1)$. For $q = 1$ formulae (A.23) and (A.22) read

$$y_{n+1} = y_n + p_1 k_1, \quad k_1 = hf(x_n, y_n).$$

Here we have only one unknown, namely p_1. To find it we use formula (A.27) for $j = 1$ which leads to

$$y'(x_n) = y'_{n+1}(0). \quad \text{(A.29)}$$

From the IVP written on the interval $[x_n, x_{n+1}]$ we obtain

$$y'(x_n) = f(x_n, y(x_n)) = f(x_n, y_n). \quad \text{(A.30)}$$

On the other hand

$$y_{n+1}(h) = y_n + p_1 k_1(h),$$

which yields

$$y'_{n+1}(h) = p_1 k'_1(h) = p_1 f(x_n, y_n).$$

For $h = 0$ we therefore have

$$y'_{n+1}(0) = p_1 f(x_n, y_n). \quad \text{(A.31)}$$

By (A.29), (A.30), and (A.31) we deduce that $p_1 = 1$ and therefore the Runge–Kutta method of order 1 for Problem (A.15) is

$$y_{n+1} = y_n + hf(x_n, y_n), \quad \text{(A.32)}$$

which is **Euler's method.**

Runge–Kutta methods of order 2 $(q = 2)$. We first write the corresponding Formula (A.23):

$$\begin{aligned}y_{n+1} &= y_n + p_1 k_1 + p_2 k_2 \\ &= y_n + p_1 hf(x_n, y_n) + p_2 hf(x_n + \alpha_2 h, y_n + b_{21}hf(x_n, y_n)).\end{aligned}$$

We develop the last term above to get the corresponding formula (A.25) which is

$$y_{n+1} = y_n + p_1 h f(x_n, y_n) + p_2 h f(x_n, y_n)$$

$$+ h^2 \left[p_2 \alpha_2 \frac{\partial f}{\partial x}(x_n, y_n) + p_2 b_{21} f(x_n, y_n) \cdot \frac{\partial f}{\partial y}(x_n, y_n) \right] + O(h^3).$$
(A.33)

The corresponding Formula (A.24) is

$$y(x_{n+1}) = y(x_n) + h y'(x_n) + \frac{h^2}{2} y''(x_n) + O(h^3)$$

$$= y_n + h f(x_n, y_n) + \frac{h^2}{2} \left[\frac{\partial f}{\partial x}(x_n, y_n) + f(x_n, y_n) \frac{\partial f}{\partial y}(x_n, y_n) \right] + O(h^3).$$
(A.34)

Comparing the terms in h from formulae (A.33) and (A.34) we get the system

$$\begin{cases} p_1 + p_2 = 1 \\[2mm] p_2 \alpha_2 = \dfrac{1}{2} \\[2mm] p_2 b_{21} = \frac{1}{2}. \end{cases}$$
(A.35)

The system above has three equations and four unknowns $(p_1, p_2, \alpha_2, b_{21})$. Hence its solution is not unique. For instance, a solution is

$$p_1 = p_2 = \frac{1}{2}, \qquad \alpha_2 = b_{21} = 1.$$

The corresponding Runge–Kutta method, namely the **Euler–Cauchy method**, is defined by

$$\begin{cases} y_{n+1} = y_n + \dfrac{1}{2}(k_1 + k_2), \\ k_1 = h f(x_n, y_n), \\ k_2 = h f(x_n + h, y_n + k_1). \end{cases}$$

Another solution of system (A.35) is

$$p_1 = 0, \quad p_2 = 1, \quad \alpha_2 = b_{21} = \frac{1}{2}.$$

The corresponding Runge–Kutta method, namely the **Euler modified method**, is therefore defined by

$$y_{n+1} = y_n + h f \left(x_n + \frac{h}{2}, y_n + \frac{h}{2} f(x_n, y_n) \right).$$

For higher-order methods the systems to be solved are more complicated and each has more than one solution. For practical reasons methods of order higher than 5 are not used. We cite here the **standard Runge–Kutta method of order 4**:

$$\begin{cases} y_{n+1} = y_n + \dfrac{1}{6}(k_1 + 2k_2 + 2k_3 + k_4) \\ k_1 = hf(x_n, y_n), \\ k_2 = hf(x_n + \dfrac{h}{2}, y_n + \dfrac{k_1}{2}) \\ k_3 = hf(x_n + \dfrac{h}{2}, y_n + \dfrac{k_2}{2}) \\ k_4 = hf(x_n + h, y_n + k_3). \end{cases} \tag{A.36}$$

The MATLAB function *ode45* uses the **Runge–Kutta–Fehlberg method** with adaptive step (h can change from a subinterval to the next one). This method is a combination of methods of order 4 and 5.

Let us remark that all RK methods have the general form

$$y_{n+1} = y_n + h\Phi(x_n, y_n, h), \tag{A.37}$$

where Φ, called *the Henrici function*, is an approximation of f on the interval $[x_n, x_{n+1}]$. For example, in the case of Euler's method above we have

$$\Phi(x_n, y_n, h) = f(x_n, y_n),$$

whereas for the Euler modified method, Φ has the form

$$\Phi(x_n, y_n, h) = f\left(x_n + \frac{h}{2}, y_n + \frac{h}{2}f(x_n, y_n)\right).$$

The convergence of Runge–Kutta methods. We consider once again Problem (A.15), where we change the notation of the initial value (to fit with the numerical method):

$$\begin{cases} y'(x) = f(x, y(x)), \\ y(x_0) = \eta, \end{cases} \tag{A.38}$$

and the general RK method written under the form (A.37), more exactly

$$\begin{cases} y_{n+1} = y_n + h\Phi(x_n, y_n, h), & n = 0, 1, \dots, N-1 \\ y_0 = \eta_h. \end{cases} \tag{A.39}$$

Here $h \in [0, h^*]$, with $h^* > 0$. We suppose that Φ is continuous.

Definition A.15. *The numerical method (A.39) is consistent with IVP (A.38) if for any solution of the corresponding ODE*

$$y'(x) = f(x, y(x)),$$

the limit of the sum

$$\sum_{n=0}^{N-1} |y(x_{n+1}) - y(x_n) - h\Phi(x_n, y(x_n), h)|$$

is 0, as $h \to 0$.

The following result holds (e.g., [CM89, Section 5.2]).

Theorem A.16. *Assume that*

$$\Phi(x, y, 0) = f(x, y) \text{ for any } (x, y).$$

Then the Runge–Kutta method (A.39) is consistent with IVP (A.38) in the sense of Definition A.15.

Definition A.17. *The numerical method (A.39) is stable if there exists a constant $\tilde{M} > 0$ independent of h such that for all sequences y_n, z_n, ε_n, $n = 0, 1, \ldots, N$ that satisfy*

$$y_{n+1} = y_n + h\Phi(x_n, y_n, h),$$

and

$$z_{n+1} = z_n + h\Phi(x_n, z_n, h) + \varepsilon_n,$$

we have

$$\max\{|z_n - y_n| \ ; \ 0 \leq n \leq N\} \leq \tilde{M}\left[|y_0 - z_0| + \sum_{n=0}^{N-1} |\varepsilon_n|\right].$$

The definition above means that "small" perturbations of the data of the numerical method imply a "small" perturbation of the numerical solution. This is very important from a practical point of view because it also takes into account the roundoff errors of the computer. The following theorem is also valid (e.g., [CM89, Section 5.2]).

Theorem A.18. *Assume that there exists a constant $\Lambda > 0$ such that*

$$|\Phi(x, y, h) - \Phi(x, z, h)| \leq \Lambda|y - z| \text{ for any } (x, y, h), (x, z, h) \in D(\Phi).$$

Then the Runge–Kutta method (A.39) is stable in the sense of Definition A.17. Moreover $\tilde{M} = e^{\Lambda c}$, where $\tilde{M}$ is the constant from Definition A.17.

Definition A.19. *The numerical method (A.39) is convergent if*

$$\lim_{h \to 0} \eta_h = \eta$$

implies

$$\lim_{h \to 0} \max\{|y(x_n) - y_n|; \ 0 \leq n \leq N\} = 0,$$

where y is the solution of (A.38) and y_n is the solution of (A.39).

Theorem A.20. *If the numerical method (A.39) is stable in the sense of Definition A.17 and consistent in the sense of Definition A.15, then it is convergent in the sense of Definition A.19.*

Proof. Let us denote $z_n = y(x_n)$ for all n, and

$$\varepsilon_n = y(x_{n+1}) - y(x_n) - h\Phi(x_n, y(x_n), h).$$

It is obvious that the sequences z_n and ε_n satisfy the condition from Definition A.17. Using Definition A.15 we get that the sum $\sum_{n=0}^{N-1} |\varepsilon_n|$ converges to 0 for $h \to 0$. Now Definition A.17 yields

$$\max\{|y(x_n) - y_n|;\ 0 \le n \le N\} \le \tilde{M}\left[|\eta_h - \eta| + \sum_{n=0}^{N-1} |\varepsilon_n|\right],$$

and therefore we get the conclusion.

As a consequence of the previous theorems we obtain the following.

Theorem A.21. *Assume that Φ satisfies the condition from Theorem A.16 and the one from Theorem A.18. Then the Runge–Kutta method (A.39) is convergent in the sense of Definition A.19.*

For more details about (the convergence of) Runge–Kutta methods we recommend [CM89, Chapter 5].

References

[Ada75]   Adams, R.A. *Sobolev Spaces*. Academic Press, New York (1975)

[All07]   Allen, L.J.S. *An Introduction to Mathematical Biology*. Pearson Prentice–Hall, Upper Saddle River, NJ (2007)

[AAA08]   Aniţa, L.-I., Aniţa, S. Arnăutu, V. Global behavior for an age-dependent population model with logistic term and periodic vital rates. *Appl. Math. Comput.* **206**, 368–379 (2008)

[Ani00]   Aniţa, S. *Analysis and Control of Age-Dependent Population Dynamics*. Kluwer Academic, Dordrecht (2000)

[AA05]   Aniţa, S. Arnăutu, V. Some aspects concerning the optimal harvesting problem. Sci. Annals Univ. "Ion Ionescu de la Brad" *Iaşi* **48**, 65–73 (2005)

[AAS09]   Aniţa, S., Arnăutu, V., Ştefănescu, R. Numerical optimal harvesting for a periodic age-structured population dynamics with logistic term. *Numer. Funct. Anal. Optim.* **30**, 183–198 (2009)

[AC02]   Aniţa, S., Capasso, V. A stabilizability problem for a reaction-diffusion system modelling a class of spatially structured epidemic model. *Nonlinear Anal. Real World Appl.* **3**, 453–464 (2002)

[AC09]   Aniţa, S., Capasso, V. A stabilization strategy for a reaction-diffusion system modelling a class of spatially structured epidemic systems (think globally, act locally). *Nonlinear Anal. Real World Appl.* **10**, 2026–2035 (2009)

[ACa09]   Aniţa, S., Capasso, V. Stabilization for a reaction-diffusion system modelling a class of spatially structured epidemic systems. The periodic case. In Sivasundaram, S., et al. (Eds.) *Advances in Dynamics and Control: Theory, Methods, and Applications*, Cambridge Scientific, Cambridge, MA (2009)

[ACv09]   Aniţa, S., Capasso, V. On the stabilization of reaction-diffusion system modelling a class of man-environment epidemics: A review. *Math. Methods Appl. Sci.* 2009; Special Issue (J. Banasiak et al. Eds.) 1–6 (2009)

[AFL09]   Aniţa, S., Fitzgibbon, W., Langlais, M. Global existence and internal stabilization for a class of predator-prey systems posed on non coincident spatial domains. *Discrete Cont. Dyn. Syst. B* **11**, 805–822 (2009)

[AIK98]   Aniţa, S., Iannelli, M., Kim, M.-Y., Park, E.-J. Optimal harvesting for periodic age-dependent population dynamics. *SIAM J. Appl. Math.* **58**, 1648–1666 (1998)

[A66]   Armijo L. Minimization of functions having Lipschitz continuous first partial derivatives. *Pac. J. Math.* **16**, 1–3 (1966)

[AN03]   Arnăutu, V., Neittaanmäki, P. *Optimal Control from Theory to Computer Programs*. Kluwer Academic, Dordrecht (2003)

[Aro85]   Aronson, D.G. The role of diffusion in mathematical population biology: Skellam revisited. In: Levin, S. (Ed.) *Mathematics in Biology and Medicine*, Springer-Verlag, Berlin (1985)

[Bar93]   Barbu, V. *Analysis and Control of Nonlinear Infinite Dimensional Systems*. Academic Press, Boston (1993)

[Bar94]   Barbu, V. *Mathematical Methods in Optimization of Differential Systems*. Kluwer Academic, Dordrecht (1994)

[Bar98]   Barbu, V. *Partial Differential Equations and Boundary Value Problems*. Kluwer Academic, Dordrecht (1998)

[BP86]   Barbu, V., Precupanu, T. *Convexity and Optimization in Banach Spaces*. D. Reidel, Dordrecht (1986)

[BC98]   Borrelli, R.L., Coleman, C.S. *Differential Equations. A Modeling Perspective*. John Wiley & Sons, New York (1998)

[Bre83]   Brezis, H. *Analyse Fonctionnelle. Théorie et Applications*. Masson, Paris (1983)

[B85]   Brokate, M. Pontryagin's principle for control problems in age-dependent population dynamics. *J. Math. Biol.* **23**, 75–101 (1985)

[BG80]   Burghes, D., Graham, A. *Introduction to Control Theory, Including Optimal Control*. Ellis Horwood Division of John Wiley & Sons, Chichester, UK (1980)

[BKL97]   Butler, S., Kirschner, D., Lenhart, S. Optimal control of chemoterapy affecting the infectivity of HIV. In: Arino, O., Axelrod, D., Kimmel, M. (Eds.) *Advances in Mathematical Population Dynamics-Molecules, Cells and Man*. World Scientific, Singapore (1997)

[C84]   Capasso, V. Asymptotic stability for an integro-differential reaction-diffusion system. *J. Math. Anal. Appl.* **103**, 575–588 (1984)

[Cap93]   Capasso, V. *Mathematical Structures of Epidemic Systems*. Lecture Notes in Biomathematics, Vol. 97, 2nd corrected printing, Springer, Heidelberg (2008)

[CW97]   Capasso, V., Wilson, R.E. Analysis of a reaction-diffusion system modelling man–environment–man epidemics. *SIAM J. Appl. Math.* **57**, 327–346 (1997)

[CLW69]   Carnahan, B., Luther, H.A., Wilkes, J.O. *Applied Numerical Methods*. John Wiley & Sons, New York (1969)

[Cea71]   Céa, J. *Optimisation. Théorie et Algorithmes*. Dunod, Paris (1971)

[Cea78]   Céa, J. *Lectures on Optimization. Theory and Algorithms*. Tata Institute of Fundamental Research, Springer-Verlag, Bombay (1978)

[Che86]   Cherruault, Y. *Mathematical Modelling in Biomedicine. Optimal Control of Biomedical Systems*. D. Reidel, Dordrecht (1986)

[Cia94]   Ciarlet, P.G. *Introduction à l'Analyse Numérique Matricielle et à l'Optimisation*, 5e tirage. Masson, Paris (1994)

[CL55]   Coddington, E.A., Levinson, N. *Theory of Ordinary Differential Equations*. McGraw-Hill, New York (1955)

[Coo98]   Cooper, J.M. *Introduction to Partial Differential Equations with MAT-LAB*. Birkhäuser, Boston (1998)

[CM89]   Crouzeix, M., Mignot, A.L. *Analyse Numérique des Équations Differentielles* , 2e édition. Masson, Paris (1989)

[CV78]   Cruceanu, Ş., Vârsan, C. *Elements of Optimal Control and Applications in Economics* (in Romanian). Editura Tehnică, Bucharest (1978)

[CIM05]   Cusulin, C., Iannelli, M., Marinoschi, G. Age-structured diffusion in a multi-layer environment. *Nonlinear Anal. R. World. Appl.* **6**, 207–223 (2005)

[DC72]   Dorn, W.S., McCracken, D.D. *Numerical Methods with FORTRAN IV. Case Studies.* John Wiley & Sons, New York (1972)

[F37]   Fisher, R.A. The wave of advance of advantageous genes. *Ann. Eugen.* **7**, 355–369 (1937)

[FL06]   Fister, K.R., Lenhart, S. Optimal harvesting in an age-structured predator-prey model. *Appl. Math. Optim.* **54**, 1–15 (2006)

[GVA06]   Genieys, S., Volpert, V., Auger, P. Pattern and waves for a model in population dynamics with nonlocal consumption of resources. *Math. Model. Nat. Phenom.* **1**, 65–82 (2006)

[GW84]   Gerald, C.F., Wheatley, P.O. *Applied Numerical Analysis*. Addison-Wesley, Reading, MA (1984)

[GS77]   Glashoff, K., Sachs, E. On theoretical and numerical aspects of the bang-bang principle. *Numer. Math.* **29**, 93–113 (1977)

[G00]   Gourley, S.A. Travelling front solutions of a nonlocal Fisher equation. *J. Math. Biol.* **41**, 272–284 (2000)

[GS81]   Gruver, W.A., Sachs, E. *Algorithmic Methods in Optimal Control.* Pitman, London (1981)

[GM81]   Gurtin, M.E., Murphy, L.F. On the optimal harvesting of age-structured populations: some simple models. *Math. Biosci.* **55**, 115–136 (1981)

[Har02]   Hartman, P. *Ordinary Differential Equations*, 2nd edition. Society for Industrial and Applied Mathematics, Boston (2002)

[HNP95]   Ho, D.D., Neumann, A.U., Perelson, A.S., Chen, W., Leonard, J.M., Markowitz, M. Rapid turnover of plasma virions and CD4 lymphocites in HIV-1 infection. *Nature* **373**, 123–126 (1995)

[HY05]   Hritonenko, N., Yatsenko, Y. Optimization of harvesting age in integral age-dependent model of population dynamics. *Math. Biosci.* **195**, 154–167 (2005)

[Ian95]   Iannelli, M. *Mathematical Theory of Age-Structured Population Dynamics.* Giardini Editori e Stampatori, Pisa (1995)

[IM09]   Iannelli, M., Marinoschi, G. Harvesting control for an age-structured population in a multi-layered habitat. *J. Optim. Theory Appl.* **142**, 107–124 (2009)

[Ist05]   Istas, J. *Mathematical Modeling for the Life Sciences.* Springer-Verlag, New York, (2005)

[KLS97]   Kirschner, D., Lenhart, S., Serbin, S. Optimizing chemotary of HIV infection: scheduling, ammounts and initiation of treatment. *J. Math. Biol.* **35**, 775–792 (1997)

[Kno81]   Knowles, G. *An Introduction to Applied Optimal Control.* Academic Press, London (1981)

[KPP37]   Kolmogorov, A.N., Petrovskii, I.G., Piskunov, N.S. Étude de l'équation de la diffusion avec croissance de la quantité de matière et son application à un problème biologique. *Bull. Univ. État Moscou, Sér. Int., Sect. A: Math et Mécan.* **1**, 1–25 (1937)

[LM67]   Lee, E.B., Markus, L. *Foundation of Optimal Control Theory.* John Willey, New York (1967)

[LW07]   Lenhart, S., Workman, J.T. *Optimal Control Applied to Biological Models.* Chapman & Hall/CRC, Boca Raton, FL (2007)

[Lio72]   Lions, J.L. *Optimal Control of Systems Governed by Partial Differential Equations.* Springer-Verlag, Berlin (1972)

[LLW04]   Luo, Z., Li, W.T., Wang, M. Optimal harvesting control problem for linear periodic age-dependent population dynamics. *Appl. Math. Comput.* **151**, 789–800 (2004)

[Mur89]   Murray, J.D. *Mathematical Biology.* Springer-Verlag, Berlin (1989)

[Oku80]   Okubo, A. *Diffusion and Ecological Problems: Mathematical Models.* Springer-Verlag, Berlin (1980)

[O86]   Okubo, A. Dynamical aspects of animal grouping: swarms, schools, flocks and herds. *Adv. Biophys.* **22**, 1–94 (1986)

[PKD93]   Perelson, A.S., Kirschner, D.E., De Boer, R. Dynamics of HIV infection of $CD4^+$ T cells. *Math. Biosci.* **114**, 81–125 (1993)

[PN02]   Perelson, A.S., Nelson, P.W. Modeling viral infections. In: Sneyd, J. (Ed.) *An Introduction to Mathematical Modeling in Physiology, Cell Biology, and Immunology.* AMS, Providence, RT (2002)

[Pol71]   Polak, E. *Computational Methods in Optimization. A Unified Approach.* Academic Press, New York (1971)

[PFT90]   Press, W.H., Flannery, B.P., Teukolsky, S.A., Vetterling, W.T. Numerical Recipes in *C. The Art of Scientific Computing.* Cambridge University Press, New York (1990)

[PW84]   Protter, M.H., Wienberger, H.F. *Maximum Principles in Differential Equations.* Springer-Verlag, New York (1984)

[QS03]   Quarteroni A., Saleri F. *Scientific Computing with MATLAB.* Springer-Verlag, Berlin (2003)

[S51]   Skellam, A.J. Random dispersal in theoretical populations. *Biometrika* **38**, 196–218 (1951)

[Ske73]   Skellam, J.G. The formulation and interpretation of mathematical models of diffusional processes in population biology. In: Bartlett, M.S., Hiorns, R.W. (Eds.) *The Mathematical Theory of the Dynamics of Biological Populations.* Academic Press, New York (1973)

[Smo83]   Smoller, J. *Shock Waves and Reaction–Diffusion Equations.* Springer-Verlag, New York (1983)

[Son98]   Sontag, E.D. *Mathematical Control Theory. Deterministic Finite Dimensional Systems*, 2nd edition. Springer-Verlag, New York (1998)

[T84]   Thieme, H. Renewal theorems for linear periodic Volterra integral equations. *J. Integral Eqs.* **7**, 253–274 (1984)

[Thi03]   Thieme, H. *Mathematics in Population Biology.* Princeton University Press, Princeton NJ and Oxford (2003)

[Tre05]   Trélat, E. *Contrôle Optimal. Théorie et Applications.* Vuibert, Paris (2005)

[Web85]   Webb, G. *Theory of Nonlinear Age-Dependent Population Dynamics.* Marcel Dekker, New York (1985)

[Y82]   Yosida, S. An optimal control problem of the prey–predator system. *Funkt. Ekvacioj* **23**, 283–293 (1982)

[You71]   Young, D.M. *Iterative Solution of Large Linear Systems.* Academic Press, New York and London (1971)

[Zwi97]   Zwillinger, D. *Handbook of Differential Equations*, 3rd edition. Academic Press, Boston (1997)

[http1]   http://www.math.hmc.edu/resources/odes/odearchitect/examples/

Index